Applied Mathematical Sciences

EDITORIAL STATEMENT

The mathematization of all sciences, the fading of traditional scientific boundaries, the impact of computer technology, the growing importance of mathematical-computer modelling and the necessity of scientific planning all create the need both in education and research for books that are introductory to and abreast of these developments.

The purpose of this series is to provide such books, suitable for the user of mathematics, the mathematician interested in applications, and the student scientist. In particular, this series will provide an outlet for material less formally presented and more anticipatory of needs than finished texts or monographs, yet of immediate interest because of the novelty of its treatment of an application or of mathematics being applied or lying close to applications.

The aim of the series is, through rapid publication in an attractive but inexpensive format, to make material of current interest widely accessible. This implies the absence of excessive generality and abstraction, and unrealistic idealization, but with quality of exposition as a goal.

Many of the books will originate out of and will stimulate the development of new undergraduate and graduate courses in the applications of mathematics. Some of the books will present introductions to new areas of research, new applications and act as signposts for new directions in the mathematical sciences. This series may serve as an intermediate stage of the publication of material which, through exposure here, will be further developed and refined and appear later in the Mathematics in Science Series of books in applied mathematics also published by Springer-Verlag and in the same spirit as this series.

MANUSCRIPTS

The Editors welcome all inquiries regarding the submission of manuscripts for the series. Final preparation of all manuscripts will take place in the editorial offices of the series in the Division of Applied Mathematics, Brown University, Providence, Rhode Island.

Published by SPRINGER SCIENCE+BUSINESS MEDIA, LLC

L. Sirovich

Techniques of Asymptotic Analysis

With 23 Illustrations

Springer Science+Business Media, LLC 1971

Lawrence Sirovich
Division of Applied Mathematics
Brown University
Providence, Rhode Island

Originally published by Springer-Verlag New York • Heidelberg • Berlin in 1971
Softcover reprint of the hardcover 1st edition 1971

Library of Congress Catalog Card Number 70-149141

ISBN 978-0-387-90022-3 ISBN 978-1-4612-6402-6 (eBook)
DOI 10.1007/ 978-1-4612-6402-6

Applied Mathematical Sciences | Volume 2

PREFACE

These notes originate from a one semester course which forms part of the "Math Methods" cycle at Brown. In the hope that these notes might prove useful for reference purposes several additional sections have been included and also a table of contents and index.

Although asymptotic analysis is now enjoying a period of great vitality, these notes do not reflect a research oriented course. The course is aimed toward people in applied mathematics, physics, engineering, etc., who have a need for asymptotic analysis in their work. The choice of subjects has been largely dictated by the likelihood of application. Also abstraction and generality have not been pursued. Technique and computation are given equal prominence with theory.

Both rigorous and formal theory is presented — very often in tandem. In practice, the means for a rigorous analysis are not always available. For this reason a goal has been the cultivation of mature formal reasoning. Therefore, during the course of lectures formal presentations gradually eclipse rigorous presentations. When this occurs, rigorous proofs are given as exercises or in the case of lengthy proofs, reference is made to the Reading List at the end.

The Reading List contains a number of books for further reading. Among these are included those books which have influenced me in the preparation of this course. Most noteworthy, in this respect, are the treatments by Friedrichs, Erdelyi, and Dieudonne. In the case of Professor Friedrichs my debt goes back to my student days when I took a course with him on asymptotic analysis. This has had a lasting effect on me in my scientific work.

Finally, I wish to acknowledge the invaluable assistance of T. H. Chong and C. Huo in the preparation of these notes. I am also indebted to Ezora Fonseca, Katherine MacDougall and Sandra Spinacci for their careful typing of my handwritten notes and to Eleanor Addison for her excellent preparation of my sketchs.

Providence, Rhode Island
December, 1970

Lawrence Sirovich

Notation:

The material is divided into three chapters and each chapter into sections. Thus, on referring, for example, to Section 3.5 we mean Chapter 3, Section 5. Equations are individually numbered in each section, for example, (3.5.1) refers to the first equation of Section 3.5. Theorems are numbered in the same way but without punctuation marks. A bracketed number such as [1] refers to a book on the Reading List, p. 300.

TABLE OF CONTENTS

PREFACE v

CHAPTER 1. ASYMPTOTIC SEQUENCES AND ASYMPTOTIC DEVELOPMENT OF A FUNCTION

1.1. Notation and Definition 1

1.2. Operations with Asymptotic Expansions 11

Asymptotic Integration 17

Differentiation 21

1.3. Some Remarks on the Use of Asymptotic Expansions . . . 24

1.4. Summation of Asymptotic Expansions 28

CHAPTER 2. THE ASYMPTOTIC DEVELOPMENT OF A FUNCTION DEFINED BY AN INTEGRAL

2.1. Elementary Analytic Methods 38

Analytic Continuation of Functions Defined by an Integral 38

Integration by Parts 40

Asymptotic Evaluation of Indefinite Integrals 44

Asymptotic Evaluation of Integrals of the Form $\int^{x} f(x,t)dt$ 54

2.2. Laplace and Fourier Transforms at Infinity 62

Watson's Lemma 65

Fourier Integrals 74

2.3. Laplace's Formula and Its Generalization 80

2.4. Kelvin's Formula and Generalizations 86

2.5. Integrals of the Type $\int_{\alpha(\underset{\sim}{x})}^{\beta(\underset{\sim}{x})} G(\underset{\sim}{x},t)dt$ 95

Generalized Laplace Formula 96

Generalized Kelvin's Formula 100

Dispersive Wave Propagation 102

2.6. Method of Steepest Descents and the Saddle Point Formula 105

Saddle Point Method for a Complex Large Parameter 115

The Complete Asymptotic Development 117

Application to Bessel Functions 122

2.7. Applications of the Saddle Point Method 126

The Airy Integral 126

A Generalization of the Airy Integral 134

2.8. Multidimensional Integrals: Part I. Laplace, Kelvin, and Related Formulas 136

2.9. Multidimensional Integrals: Part II. Many Parameters 148

Complete Asymptotic Development 155

2.10. Asymptotic Evaluation of Integrals Involving Non-Uniformities 164

Integrals Containing a Global Maximum Near an Endpoint 165

Neighboring Saddle Points 170

2.11. Miscellaneous 176

Laplace Transforms in the Neighborhood of the Origin 176

Fourier Transforms at the Origin 181

Bromwich Integrals at Infinity 182

Bromwich Integrals at the Origin 186

CHAPTER 3. LINEAR ORDINARY DIFFERENTIAL EQUATIONS

3.0. Introduction 189

3.1. Some Topics in Matrix Analysis 192

Applications to Ordinary Differential Equations . . . 195

3.2. Matrix Theory - Continued 201

Functions of Matrices 207

Dunford-Taylor Integral 210

Construction of a Function of a Matrix 211

3.3. Linear Ordinary Differential Equations with Constant Coefficients 219

3.4. Classification and General Properties of Ordinary Differential Equations in the Neighborhood of Singular Points 226

Circuit Relations 230

Singular Points of an nth Order Scaler Ordinary Differential Equations 232

Solutions in the Neighborhood of Infinity 234

The Equation $z \frac{d\tilde{X}}{dz} = \underset{\sim}{A}_0 \underset{\sim}{X}$ 235

3.5. Linear Ordinary Differential Equations with Regular Singular Points 240

The Case of a Scaler Ordinary Differential Equation with a Regular Singular Point 254

Method of Frōbenius 256

3.6. Irregular Singular Points 259

Scaler Ordinary Differential Equations 278
Second Order Equations 284
3.7. Ordinary Differential Equations Containing a Large Parameter 289
Formal Solution 289
Turning or Transition Points 292
Connection Formulas 294
Langer's Uniform Method 296

READING LIST 300

INDEX 301

CHAPTER 1

ASYMPTOTIC SEQUENCES AND THE ASYMPTOTIC DEVELOPMENT OF A FUNCTION

1.1. Notation and Definition.

It is often the case that we desire to approximate a function when a parameter, index or independent variable tends to a specific value. For example

$$e^{-x} \approx 1 - x \quad \text{when} \quad x \quad \text{is small}$$

$$\sin(x+\varepsilon) \approx \sin x + \varepsilon \cos x, \quad \text{when} \quad \varepsilon \quad \text{is small}$$

$$J_n(x) \approx \frac{1}{(2\pi n)^{\frac{1}{2}}} \left(\frac{ex}{2n}\right)^n, \quad \text{when} \quad n \quad \text{is large.}$$

In order for such approximations to have any value it will be necessary in each case to know the domain of validity; and of more practical significance an estimate of the error in the approximation.

Aside from the additional generality, it will be convenient to perform our calculations in the complex plane.

<u>Notation.</u> Representing a point in the complex plane by $z = x + iy$, the value of a function f at the point z is denoted by $f(z)$. This does not convey the idea that $f(z)$ is analytic. For, in general, one may write

$$\tilde{f}(x,y) = \tilde{f}\left(\frac{z+\bar{z}}{2}, \frac{z-\bar{z}}{2i}\right).$$

And since $\bar{z}$, itself, is a function of z,

$$\tilde{f}(x,y) = f(z).$$

<u>Definition.</u> The letter S will be employed for a region in the complex plane. For the most part S will represent a sector. More specifically, we will write

(1.1.1) $S_R(\alpha,\beta)$: $[z| \; S_R(\alpha,\beta): 0 < |z| < R, \; \alpha < \arg z < \beta]$.

If $\alpha = -\beta$ we will write simply $S_{R\beta}$, also if $R = \infty$ we will frequently drop it as a subscript, i.e., $S_{\infty\beta} = S_\beta$.

Asymptotic Power Series.

The following is the fundamental definition of asymptotic analysis.

Definition (Asymptotic Power Series APS). Let $f(z)$ be defined in S, $\bar{S} \supset z = 0$. (Bar denotes the closure.) $f(z)$ is said to have an asymptotic power series (APS) representation of order N, given by $\sum_{i=0}^{N} a_i z^i$ if for all $n \leq N$ and arbitrary $\varepsilon > 0$, there exists $\delta > 0$ such that

$$|f(z) - \sum_{i=0}^{n} a_i z^i| < \varepsilon |z|^n, \quad z \in S$$

and $|z| < \delta(\varepsilon,n)$. Or equivalently,

$$\lim_{z \to 0} \left| \frac{f(z) - \sum_{i=0}^{n} a_i z^i}{z^n} \right| = 0, \; z \in S. \tag{1.1.2}$$

We write this as

$$f(z) \sim \sum_{i=0}^{N} a_i z^i, \quad z \in S,$$

or in terms of an equality

$$\text{APS } f(z) = \sum_{i=0}^{N} a_i z^i.$$

The above definition of an asymptotic expansion can be applied at arbitrary points of the plane by a translation and in the neighborhood of infinity by writing $1/z$ for z.

As an example consider the function

$$f(z) = e^{-\frac{1}{z^2}} + \sin z,$$

in the neighborhood of the origin. This has an essential singularity at the origin. According to the definition of APS we have

$$f(z) \sim \sum_{j=o}^{\infty} \frac{(-)^j z^{(2j+1)}}{(2j+1)!}, \quad z \in S_{\pi/4} .$$

Furthermore, we can also write

$$e^{\frac{1}{z^2}} f(z) \sim 1, \quad z \in S_{\infty}(\frac{\pi}{4}, \frac{3\pi}{4})$$

to all orders.

We see from this example that the analytic behavior of the asymptotic development of $f(z)$ changes across the rays $\theta = \frac{\pi}{4}, \frac{3\pi}{4}, \frac{5\pi}{4}, \ldots$. This phenomena is encountered frequently in asymptotic analysis.

Definition (Stokes Line). Let T represent a line in the complex plane, and let us write the operation of taking the analytic continuation across T by AC. Then if

$$AC(APS\ f(z)) \neq APS(AC\ f(z))$$

across T, then T is said to be a Stokes line of f.

Thus the rays $\theta = \frac{\pi}{4}, \frac{3\pi}{4}, \ldots$ are Stokes lines of e^{-1/z^2}.

It is of course the case that $APS(AC\ f(z))$ is the correct operation to employ.

For differentiable functions, an immediate result on the APS of a function is given by Taylor's theorem: If $f(x) \in C^n$, $x \in [a,a+h]$ then

$$f(a+h) = \sum_{m=0}^{n-1} \frac{h^m f^m(a)}{m!} + R_n$$

$$R_n = \frac{h^n}{(n-1)!} (1-\theta)^{n-1} f^{(n)}(a+\theta h), \qquad 0 \leq \theta \leq 1.$$

Often an APS is divergent. For example consider the function

$$G(x) = \int_0^\infty \frac{e^{-t}}{1+xt}\, dt \tag{1.1.3}$$

for $x > 0$ and small. Then since

$$\frac{1}{1+xt} = 1 - xt + x^2t^2 + \cdots + \frac{(-xt)^k}{1+xt}$$

we have

$$G(x) = \sum_{i=0}^{k} \int_0^\infty e^{-t}(-xt)^i dt + \int_0^\infty \frac{e^{-t}(-xt)^{k+1}}{1+xt}\, dt.$$

Then since $\int_0^\infty e^{-t}t^n dt = n!$, we have

$$\left|G(x) - \sum_{i=0}^{k} (-)^i x^i i!\right| \leq x^{k+1}(k+1)! \; .$$

Therefore, $G(x)$ has the asymptotic development $\sum_{n=0}^{N} (-1)^n n! x^n$ for all N. We write this as

$$G(x) \sim \sum_{n=0}^{\infty} (-1)^n n! x^n = \text{APS } G(x).$$

The middle summation is of course meaningless in the ordinary sense since the summation converges for no value of $x \neq 0$. We, nevertheless, find it convenient to retain this form as a notation. Thus we write

$$f(z) \sim \sum_{n=0}^{\infty} a_n z^n$$

with the meaning that $f(z)$ has for an APS the summation $\sum a_n z^n$ to all orders - whether or not the infinite sum converges.

This example illustrates clearly that an APS may be divergent. Of course it may be convergent - as would be the case for a function analytic at the point in question. Moreover, even if an APS is convergent it need not converge to the function for which it is an APS. We have already encountered this with $e^{-1/z^2} + \sin z \sim \sum (-)^j z^{2j+1}/(2j+1)! = \sin z,\ z \in S_{\pi/4}$.

In the following we will write simply

$$f \sim \sum a_i z^i$$

(without fixed limits) if a result does not hinge on whether the limit of the APS is finite or not.

A surprising amount of simplification is obtained through the introduction of the Landau symbols O and o

Definition. Let ϕ and ψ be defined in a common region S of the complex plane. Then we write

$$\phi = O(\psi)$$

if there exists an $A > 0$ such that $|\phi| \leq A|\psi|$ for all $z \in S$.

A somewhat different use of the same symbol occurs in the following definition.

Definition. Let ϕ and ψ be defined in a common region S and $z_0 \in \bar{S}$, then

$$\phi = O(\psi) \quad \text{as} \quad z \to z_0, \quad \text{in} \quad S$$

if

$$\lim_{z \to z_0, z \in S} |\phi/\psi| < \infty.$$

Another related symbol is contained in the next definition.

Definition. ϕ and ψ defined in S, $z_0 \in \overline{S}$, then

$$\phi = o(\psi) \quad \text{as} \quad z \to z_0 \quad \text{in} \quad S$$

if

$$\lim_{z \to z_0, z \in S} |\phi/\psi| = 0.$$

The two Landau symbols o and O may be compared;

If as $z \to o$, $f(z) = o(z^n)$ then also $f = O(z^n)$

If as $z \to o$, $f(z) = O(z^{n+1})$ then also $f(z) = o(z^n)$.

However, in each case the first representation is the sharper one.

Exercise 1. Use the Landau symbols to describe the following functions at infinity,

(a) $\sin z$, $0 \leq \arg z < 2\pi$

(b) $t^x + x$, x real, t a complex constant

(c) $[1+\ln z]^{1/3}/[1+z^2]^{1/2}$, $|\arg z| < \frac{\pi}{2}$.

(d) $\dfrac{1}{e^{\sqrt{z}}+e^{z^{1/3}}}$, $|\arg z| < \pi$

Exercise 2. For ϕ,ψ,χ defined in S prove

(a) If $\phi = O(\psi)$ and $\alpha > 0$ then $\phi^{\alpha} = O(\psi^{\alpha})$

Is this true if $\alpha < 0$?

(b) $\phi = O(\psi)$, $\psi = O(\chi) \Rightarrow \phi = O(\chi)$.

An important use of the Landau symbols is that it permits us to replace the $\sim$ sign by equality. For example, if

$$f \sim \sum_{i=0}^{N} a_i z^i$$

in S, then

$$\lim_{z \to 0, z \in S} |(f - \sum_{i=0}^{N} a_i z^i)/z^N| = 0.$$

Then by the definition of little o, we can write

$$f = \sum_{i=0}^{N} a_i z^i + o(z^N)$$

or if

$$f \sim \sum_{i=0}^{\infty} a_i z^i$$

as $z \to 0$, then

$$f = \sum_{i=0}^{k-1} a_i z^i + O(z^k)$$

for all k.

Asymptotic Sequences and Developments.

Until this point we have considered the behavior of a function, say in the neighborhood of the origin or at ∞, purely in terms of power series or inverse power series. I.e., we have used the monomials $z^{\pm p}$, p integer to "gauge" the behavior of the function. From experience it is clear that this is too crude for many purposes and we should permit additional gauge functions in our treatment, e^z, z^λ (λ a complex constant), ln z, and so on. Such gauges are necessary if we are to accurately describe the asymptotic behavior of a wide variety of functions. We introduce arbitrary gauge functions through the definition of asymptotic sequences, AS.

Definition (Asymptotic Sequences, AS). A sequence of functions $\{\phi_n\}$ all defined in a common domain S is said to be an AS as $z \to z_0$, $z \in S$ if

$$\phi_{n+1} = o(\phi_n) \quad \text{as} \quad z \to z_0.$$

Thus $\{z^n\}$ and $\{z^{-n}\}$ for integer are AS at the origin and at infinity, respectively. Other examples of AS are: $\{e^{-nz}\}$ $z \to \infty$, $\arg z < \frac{\pi}{2} - \varepsilon$, $\varepsilon > 0$, $\{(\ln z)^{-n}\}$ $z \to \infty$, $\{e^{-x^n}\}$, x real $\to \infty$.

Exercise 3. Let $\{\lambda_n\}$ represent a sequence of complex constants such that

$$\text{Re}\,\lambda_{n+1} > \text{Re}\,\lambda_n.$$

Demonstrate that $\{z^{-\lambda_n}\}$ is an AS as $z \to \infty$. Is there a restriction on the sector of validity?

Having introduced general gauge functions by means of AS we can also generalize the definition of an asymptotic development AD.

Definition (General AD). Let $f(z)$, $\{\phi_n(z)\}$ be defined in S and $z_0 \in \overline{S}$, and $\{\phi_n\}$ be an AS as $z \to z_0$. Then f(z) is said to have an AD in $\{\phi_n\}$ as $z \to z_0$ of order N if there exist constants c_k such that

$$f = \sum_{k=0}^{N} c_k\phi_k + o(\phi_N).$$

It is direct that if f has an AD of order N it has an AD for all $K \leq N$.

From the existence of an AD for a function of z in terms of an AS $\{\phi_n\}$ the determination of the constants c_k is direct. For we can successively determine the constants c_k through the formula,

$$c_k = \lim_{z \to z_0, z \in S} [(f - \sum_{n=0}^{k-1} c_i \phi_i)/\phi_k].$$

This construction along with the above definition constitutes the idea of asymptotic development in the sense of Poincaré.

The notion of an asymptotic development in the sense of Poincaré is not the only such useful idea. A more general definition is illustrated by the following example.

A function $f(x)$ sufficiently smooth on some interval of the real axis may be represented there by a series of Bessel functions,

$$f(x) = \sum_{i=0}^{\infty} a_i J_i(x)$$

where the a_i are constants. In the neighborhood of the origin $\{J_i(x)\}$ forms an AS. Then since $J_i(x) = O(x^i)$ as $x \to 0$ we can say

$$f(x) = \sum_{i=1}^{N} a_i J_i(x) + o(x^N).$$

In this form it is seen that although $\{J_n(x)\}$ is the AS used in the AD, the AS $\{x^n\}$ is used as the gauge. This particular example is somewhat trivial in that $\{J_n(x)\}$ itself may be used as the gauge since $J_i(x) = O(x^i)$ in the neighborhood of the origin.

More generally, one says that f has an AD of order N, $f \sim \sum f_n$ with respect to an AS $\{\phi_n\}$ $z \to z_0$ if

$$f = \sum^{k} f_n + o(\phi_k) \quad \text{for each successive} \quad k \leq N.$$

We will not find enough use of this idea to pursue it further. (For more details see, Erdelyi, [2] and Archive for Rat. Mech. Anal. 7, 1(1961)).

Extended Asymptotic Development.

Consider

$$\sum_{n=1}^{\infty} \frac{\sin nx}{x^n} .$$

For $x > 1$ this series is convergent. However, according to our definitions this series is not asymptotic as $x \to \infty$, due to the fact that $\{x^{-n} \sin nx\}$ is not an AS. This is only a minor problem and may be circumvented as follows (see also Dieudonné [13]).

Let $\{\phi_n(z)\}$ be an AS in a sector S as $z \to \infty$. Let C represent a class of functions $C(z)$ defined in S and having the following properties in S.

(1) All $C(z)$ are bounded as $|z| \to \infty$ in S.

(2) No function of C aside from 0 vanishes as $|z| \to \infty$ in S.

(3) The linear combination of functions of C taken with complex constants also belongs to C.

Definition. We then say that a function $f(z)$ defined in S has an AD, as $|z| \to \infty$ in S, in the extended sense if

$$f = \sum_{i=0}^{m} c_i(z)\phi_i(z) + o(\phi_m(z))$$

where the $c_i \in C$.

Regarding this definition we see that it allows us to approximate a function by means of series with a controlled small remainder - and this is the crucial property. Another property of interest is uniqueness which is immediate, for suppose

$$f \sim \sum_{i=0} c_i(z)\phi_i(z)$$

$$f \sim \sum_{i=0} \tilde{c}_i(z)\phi_i(z)$$

for $|z| \to \infty$ $z \in S$. Then by subtraction to leading order we have

$$\{c_0(z) - \tilde{c}_0(z)\}\phi_o = o(\phi_o)$$

or

$$c_0(z) - \tilde{c}_0(z) = o(1).$$

But since $\{c_0(z) - \tilde{c}_0(z)\} \in C$ by (3), we violate (2) unless

$$c_0(z) = \tilde{c}_0(z)$$

and so forth.

One easily verifies that cos px, sin qx, for real p and q belong to C for the real line.

In the following, we will not generally make a distinction between an AD and an AD in the extended sense.

1.2. Operations With Asymptotic Expansions.

In this section we first deal with a number of elementary manipulations of APS, including the sum, difference and product of the APS of functions. A precondition for such manipulations is that the various functions have a common region of definition including limit points in which their APS are valid. In order to avoid tiresome repetition in the statement of theorems below we henceforth assume that all functions and their APS have been properly defined. Also we point out again that there is no loss in generality in taking the origin as the limit point, hence we shall assume $\overline{S} \supset z = 0$.

Theorem 120. The linear combination of APS is the APS of the correspondingly combined functions. Symbolically

$$\text{APS}(af(x) + bg(z)) = a \text{ APS } f + b \text{ APS } g.$$

Proof. For z in the common domain of definition of f and g we have,

$$f(z) = \sum_{n=o}^{N} a_n z^n + o(z^N)$$

$$g(z) = \sum_{n=o}^{M} b_n z^n + o(z^M).$$

Take $P = \min(N,M)$, then

$$af + bg = a \sum_{n=o}^{P} a_n z^n + b \sum_{n=o}^{P} b_n z^n + o(z^P).$$

Moreover, this is true for all indices $Q \leq P$.

Theorem 121. The APS of a function is unique.

Proof. Suppose $f(z)$ has the two APS for $z \in S$,

$$f \sim \sum a_i z^i$$

$$f \sim \sum b_i z^i.$$

Then from the previous theorem

$$0 \sim \sum (a_i - b_i) z^i$$

and from the Poincaré construction

$$a_i = b_i \,, \qquad i = 0,1,\ldots \quad \text{successively.}$$

It should be understood that the theorem demonstrates that $f(z)$ has a unique AD in terms of the AS $\{z^n\}$. A given function will have a different AD for each AS. For example, for $z \to \infty$, we have

$$\frac{1}{1+z} \sim \sum_{n=1}^{\infty} \frac{(-)^{n-1}}{z^n} \,, \qquad \text{in } \{z^{-n}\}$$

$$\frac{1}{1+z} \sim \sum_{n=1}^{\infty} \frac{(z-1)}{z^{2n}} \,, \qquad \text{in } \{\frac{(z-1)}{z^{2n}}\}$$

$$\frac{1}{1+z} \sim \sum_{n=1}^{\infty} (-)^{n-1} \frac{(z^2-z+1)}{z^{3n}} \,, \qquad \text{in } \{\frac{z^2-z+1}{z^{3n}}\} \,.$$

Corollary. The AD of a function in an AS is unique.

In fact the above and the previous theorems go through without change in proof for asymptotic expansions in general AS.

The converse of the last theorem is false. An asymptotic expansion does not uniquely determine a function. For example suppose as $z \to 0$

$$f(z) \sim \sum a_i z^i \,, \quad z \in S_{\pi/2-\delta} \quad (\delta > 0 \text{ arbitrarily small})$$

[Since $S_{\pi/2-\delta}$ often occurs, we will frequently adopt the shorthand $S^\delta = S_{\pi/2-\delta}$.]

Then also

$$f(z) + e^{-1/z} \sim \sum a_i z^i \,, \quad z \in S^\delta .$$

Theorem 122. The formal multiplication of APS is the APS of the product of the corresponding functions. Symbolically, if

$$\left.\begin{aligned} f(z) &\sim \sum a_i z^i \\ g(z) &\sim \sum b_i z^i \end{aligned}\right\} \quad z \in S$$

then

$$f(z)g(z) \sim \sum c_i z^i$$

with

$$c_i = \sum_{p+q=i} a_p b_q$$

<u>Proof</u>. Write

$$f = \sum_{i=0}^{n} a_i z^i + O(z^{n+1}) = f_n + O(z^{n+1})$$

and similarly

$$g = g_n + O(z^{n+1}).$$

Then

$$\begin{aligned} fg &= f_n g_n + (f-f_n)g_n + f(g-g_n) \\ &= f_n g_n + O(z^{n+1}) \\ &= \sum_{i=0}^{n} c_i z^i + (f_n g_n - \sum_{i=1}^{n} c_i z^i) + O(z^{n+1}) \\ &= \sum_{i=1}^{n} c_i z^i + O(z^{n+1}). \end{aligned}$$

Theorem 123. If $f \sim \sum a_i z^i$, $z \in S$ and $\lim_{z \to 0, z \in S} f(z) \neq 0$ then f^{-1} has an APS and moreover it is given by the formal inversion of $\sum a_i z^i$.

Proof. To begin with we construct the formal inverse. First write

$$f_n = \sum_{i=0}^{n} a_i z^i$$

and by hypothesis $a_0 \neq 0$. Next, construct a g_n

$$g_n = \sum_{i=0}^{n} b_i z^i$$

such that

$$\lim_{z \to 0} z^{-k}[1-g_n f_n] = 0, \quad k = 0,1,\ldots,n.$$

Then since $f_n g_n = a_0 b_0 + (a_1 b_0 + a_0 b_1)z + \cdots + \sum_{i+j=k} a_i b_j z^k + \cdots$ we have

$$b_0 = \frac{1}{a_0}, \quad b_1 = -\frac{a_1}{a_0^2}, \quad b_2 = -\frac{a_2}{a_0^2} + \frac{a_1^2}{a_0^3}, \ldots .$$

Finally write

$$(f^{-1}-g_n) = f^{-1}(1-fg_n) = f^{-1}[1-f_n g_n + g_n(f_n - f)] = O(z^{n+1}).$$

Another result of the same type is given in the following:

Exercise 4. Prove the following

Theorem 124. If $g(w)$ is defined in $\mathcal{S}$, $\overline{\mathcal{S}} \supset w = 0$, and has an APS there,

$$g \sim \sum_{i=0} b_i w^i.$$

and if $f(z)$ defined in S, such that

$$f \sim \sum_{i=1} a_i z^i$$

and if for $z \in S$, $f(z) \in \mathscr{S}$.

Then

$$g(f(z)) \sim \sum_{i=o} c_i z^i$$

where the c_i are found by the formal procedure. (In particular, find c_o, c_1, c_2.)

The above three theorems hold also for asymptotic expansions in the AS $\{z^{-n}\}$, n integer. This is seen directly by substituting z^{-1} for z and allowing $z \to \infty$. However, unlike Theorems 120 and 121, Theorems 122, 123, 124 have no general validity. For example consider the AS $\{z^n \ell nz\}$ at the origin. Multiplying the members of the sequence or reciprocating them leads to terms not in the sequence. Such difficulties can often be avoided by trivial manipulation. Suppose for example,

$$f(z) \sim \sum_{i=1} a_i z^i \ell nz$$

then by considering $f(z)/\ell nz$ the series can be reciprocated according to Theorem 123. Or on considering two such expansions we can by the same device multiply them by Theorem 122.

Although we cannot always so easily avoid these difficulties, it should be mentioned that such problems are more technical than real. Thus for example by multiplying two entire different AD's we really get an asymptotic approximation (only the bookkeeping may be difficult) but not necessarily in the class of either of the associated AS. We can actually avoid such problems altogether by considering a sufficiently large class of gauge function, say e.g., consider the class

(1.2.1) $$\{x^p(\ln x)^q e^{P(x)}\}$$

with

$$P(x) = \sum_{i=o}^{N} \alpha_i x^{\gamma_i}, \quad \gamma_o \geq \gamma_1 \cdots > \gamma_N > 0$$

generated by considering all constants p,q and forms $P(x)$. However, now other "technical" problems appear, especially on going to the complex plane. We will not pursue this further, but rather settle for the simplicity of APS. The main thrust of these remarks is that in any particular situation one should not be handcuffed by the relatively narrow theorems we are proving - but rather use them as a guide in extended situations.

Exercise 5. Given that in some S, and for $z \to 0$

$$g(z) \sim \sum_n a_n(e^{-z^n}-1)$$

and

$$f(z) \sim \sum_n b_n \frac{z^n}{\ln z}$$

find an AD for fg. Prove your result.

We complete this section with some results on the integration and differentiation of asymptotic expansions.

Asymptotic Integration.

Theorem 125. $f(z)$ defined in S and such that

$$f(z) = \sum_{i=o}^{N} a_i z^i + o(z^N).$$

Then $\int_{0}^{z} f(z)dz$, where the path is a ray from the origin, has an APS expansion of order $N + 1$.

Proof. By hypothesis

$$f(z) - \sum_{i=0}^{N} a_i z^i = o(z^N)$$

and on integrating along the ray to z and taking absolute values

$$\left|\int_{0}^{z} f(z)dz - \sum_{i=0}^{N} \frac{a_i z^{i+1}}{i+1}\right| \leq \int_{0}^{z} \varepsilon |z|^N \, |dz| = o(z^{N+1}).$$

Clearly the same result holds for any path of integration lying in S and of length $O(|z|)$.

Corollary. If $f(z)$, defined in S, has a derivative, and furthermore

$$f' = \sum^{N} a_i z^i + o(z^N)$$

then $f(z)$ has an APS and its differentiation gives the APS of f'.

This follows trivially from integrating and then differentiating the above development of f'.

Strictly speaking Theorem 125 does not apply to the AS $\{z^{-n}\}$, $n = 0,1,\ldots,$ at ∞. That is if

$$f(z) = \sum_{i=0}^{N} \frac{a_i}{z^i} + o(z^{-N})$$

the term by term integration fails due to the terms

$$\int_{z}^{\infty} a_0 dz, \quad \int_{z}^{\infty} \frac{a_1}{z} dz.$$

In such situations it is often useful to consider instead $f - a_o - \frac{a_1}{z}$ which presents no problem. For the case of integrals at infinity it is useful for us to obtain some additional results relating the integration of an AD. We first remark that depending on whether

$$\int_a^\infty f(z)dz$$

converges or not one is in general interested in the two types of indefinite integrals

$$\int_a^z f(z)dz, \quad \int_z^\infty f(z)dz.$$

Also it is a matter of practice that the integrals such as these have their paths of integration along straight lines in the neighborhood of ∞. As a practical matter therefore we assume that the above integrals have straight line paths. Finally, under simple transformations these integrals can generally be reduced to integrals along the real line. For example, consider

$$\int_z^\infty f(\tilde{z})d\tilde{z}$$

with the path of integration given by

$$\text{Im } z = y = \text{constant}.$$

The integral may then be written as

$$\int_x^\infty f(\tilde{x}+iy)d\tilde{x}.$$

We, therefore, consider integrals of the form

$$\int_x^{\infty} f(s)ds, \int_a^x f(s)ds$$

for x large and f in general complex.

<u>Theorem 126</u>. For $f(s)$ complex and $g(s) > 0$, both piecewise continuous on (a,∞) as $x \to \infty$

<u>Part A</u>. If $\int_a^{\infty} g(s)ds = \infty$, then

<u>A1</u> $$f = O(g) \Rightarrow \int_a^x f(s)ds = O(\int_a^x g(s)ds)$$

<u>A2</u> $$f = o(g) \Rightarrow \int_a^x f(s)ds = o(\int_a^x g(s)ds)$$

<u>A3</u> $$f \sim cg, |c| \neq 0 \Rightarrow \int_a^x f(s)ds \sim c \int_a^x g(s)ds.$$

<u>Part B</u>. If $\int_a^{\infty} g(s)ds < \infty$, then

<u>B1</u> $$f = O(g) \Rightarrow \int_x^{\infty} f(s)ds = O(\int_x^{\infty} g(s)ds)$$

<u>B2</u> $$f = o(g) \Rightarrow \int_x^{\infty} f(s)ds = o(\int_x^{\infty} g(s)ds)$$

<u>B3</u> $$f \sim cg, |c| \neq 0 \Rightarrow \int_x^{\infty} f(s)ds \sim c \int_x^{\infty} g(s)ds.$$

<u>Proofs</u> B1. For x sufficiently large there exist κ such that $|f(s)| \leq \kappa g(s)$, $s \geq x$ and

$$\left|\int_x^{\infty} f(s)ds\right| \leq \int_x^{\infty} |f(s)|ds \leq \kappa \int_x^{\infty} g(s)ds.$$

<u>B2</u>. For any small $\varepsilon > 0$ there exists an x large such that $|f(s)| \leq \varepsilon g(s)$, $s > x$

and therefore,

$$\left|\int_x^\infty f(s)ds\right| \leq \int_x^\infty |f(s)|ds \leq \varepsilon \int_x^\infty g(s)ds.$$

B3. Consider $f - cg = o(g)$ and apply B2.

A1. There exists K perhaps very large such that

$|f(s)| \leq Kg(s), \ s \geq a$

and therefore,

$$\left|\int_a^x f(s)ds\right| < K \int_a^x g(s)ds.$$

A2. For small $\varepsilon > 0$ there exists x_0 such that

$|f(s)| \leq \varepsilon g(s), \ s \geq x_0, x_0 \geq a$

and hence

$$\left|\int_{x_0}^x f(s)ds\right| \leq \varepsilon \int_{x_0}^x g(s)ds \leq \varepsilon \int_a^x g(s)ds.$$

Since $\int_a^\infty g(s)ds$ diverges choose x sufficiently large so that

$$\varepsilon \int_a^x g(s)ds \geq \left|\int_a^{x_0} f(s)ds\right|$$

and adding to above gives the proof.

A3. Apply A2 to $f - cg = o(g)$.

Differentiation of APS.

As the corollary to Theorem 125 already indicates statements about the differentiation of an AD are weak. In general, the differentiation of an AD is not the AD of the differentiated function. E.g. consider

$$f(z) = e^{-1/z}\sin(e^{1/z})$$

for $z \in S^\delta (= S_{\pi/2-\delta}, \delta > 0$, arbitrarily small$)$ $z \to \infty$. Then $f \sim 0$ to all orders. On the other hand

$$f'(z) = z^{-2}e^{-1/z}\sin(e^{1/z}) - z^{-2}\cos(e^{1/z})$$

which has a non-trivial APS at ∞. Stronger results are available if we admit analyticity. If the function is analytic in the neighborhood of the limit point the case becomes trivial, i.e.,

<u>Theorem 127</u>. $f(z)$ analytic for $0 < |z| < R$ and

$$f(z) \sim \sum a_i z^i$$

then the APS converges.

<u>Proof</u>. Since $\lim_{z \to 0} f(z) = a_0$ exists the origin is a removable singularity of f and f is, therefore, analytic at $z = 0$. Hence it has a convergent power series expansion at the origin. By the uniqueness theorem (Theorem 121) this is identical to $\sum_{i=0}^{\infty} a_i z^i$.

The term by term differentiation then follows from function theory.

If the limit point does not turn out to be a removable singularity this result of course will not hold. E.g., for $z \approx 0$

$$\int_0^\infty \frac{e^{-t}dt}{1+zt} \sim \sum (-)^i i! z^i \quad \text{for } |\arg z| < \frac{3\pi}{2}$$

(will be shown later). In this case the origin is a branch point. Nevertheless, the term by term differentiation of the APS is still permitted. This result is is contained in the following.

Theorem 128. $f(z)$ analytic for $z \in S_R(\alpha,\beta)$ and in that sector

$$f(z) \sim \sum a_i z^i$$

then

$$f'(z) \sim \sum a_i i z^{i-1}$$

for $z \in S_R(\alpha+\delta,\beta-\delta)$, $\delta > 0$.

Proof. Since $f(z)$ is analytic in S we may write

$$f(z) = \sum_{i=0}^{n} a_i z^i + \epsilon_n(z) z^n .$$

About $\epsilon_n(z)$ we can say that it is analytic in S and that $\epsilon_n(z) \to 0$ as $z \to 0$ in S and hence $\epsilon_n(z)$ is uniformly continuous in $\overline{S}_R(\alpha,\beta)$.

Differentiating the expression for $f(z)$

$$f'(z) = \sum_{i=1}^{n} i a_i z^{i-1} + \epsilon_n(z) n z^{n-1} + \epsilon_n'(z) z^n .$$

Using Cauchy's theorem we can write

$$\epsilon_n'(z) = \frac{1}{2\pi i} \oint_c \frac{\epsilon_n(\zeta) d\zeta}{(\zeta - z)^2}$$

where the closed contour c lies entirely in S. In particular, we choose c to be a circle of radius $|z|\delta$ having its center in $S_R(\alpha+\beta,\beta-\delta)$ (see Fig. 1.2.1). c, therefore, lies entirely in $S_R(\alpha,\beta)$.

Denoting the maximum of $|\epsilon_n(z)|$ on such a circle around z by $M_n(z)$ we have

$$|\epsilon_n'(z)| \leq \frac{M_n(z)}{|z|\delta} .$$

And therefore

$$|f' - \sum_{i=1}^{n} a_i iz^{i-1}| \leq |z|^{n-1}\left(n|\epsilon_n(z)| + \frac{M_n(z)}{\delta}\right).$$

Which since $M_n, \epsilon_n = o(z)$ completes the proof.

Exercise 6. What is the AD of

$$f(z) = z^2 + \sin z$$

for $z \to \infty$, $|\arg z| < \pi$.

[Note this is an example of a function which has a finite AD.]

Exercise 7. For x real and positive find the AD of

$$(1+x)^{1/x}$$

for $x \approx 0$ and $x \approx \infty$.

1.3. Some Remarks on the Use of Asymptotic Expansions.

We have considered the function (see (1.1.3))

$$G(x) = \int_0^\infty \frac{e^{-t}}{1+xt}\, dt$$

for real positive x and found for $x \to 0$,

$$G = \sum_{k=0}^{N} (-)^k x^k k! + R_{N+1}(x)$$

with the specific error bound

$$|R_N(x)| \leq N! \; x^N$$

We note the following features of this AD;

(i) The error bound is measured by the first neglected term specifically the error is less than or equal to the first neglected term.

Considering the magnitude of the ratio of two successive terms, we find,

$$\frac{x^k k!}{x^{k-1}(k-1)!} = xk.$$

(ii) The terms first decrease (since by assumption $0 < x \ll 1$) and then increase (when $k \geq \frac{1}{x}$).

(iii) It follows from (i) and (ii) that for a given value of x there exists a best approximation.

Or in other words,

(iv) For a fixed value of x only a definite accuracy can be achieved.

If for a general AD say

$$f(z) \sim \sum a_i \phi_i(z)$$

in some AS $\{\phi_i\}$, the properties (i) and (ii) are known to hold then the best approximation of this development for some value z_o, say, is gotten by taking as an approximation

$$\sum_{i=o}^{N} a_i \phi_i(z_o)$$

with N chosen such that the magnitude of the first neglected term is the minimal term of series.

For example, consider the following asymptotic development for the Gamma function

$$\Gamma(x+1) = x! = \int_o^\infty e^{-t} t^x dt$$

for x large,

$$\ell n(x!) \sim \frac{1}{2}\,\ell n(2\pi) + (x + \frac{1}{x})\ell nx - x + \frac{1}{12x} - \frac{1}{360x^3}$$

$$+ \frac{1}{1260x^5} - \frac{1}{1680x^7} + \frac{1}{1188x^9} + \cdots .$$

Then using this to calculate $\ell n(1)!(=0)$ we see that the best estimate is obtained by halting the series with $\frac{1}{1260(1)^5}$. This gives

$$\ell n(1)! \sim .0005,$$

and if we assume (i) then $R_n \approx .0005$.

The properties (i) - (iv) are found to hold for most asymptotic expansions which occur in practice. In fact, one often encounters the sentiment that these effects are always valid, however, this is not true. It is easy to contrive examples for which these general effects are not valid, e.g., $x^{100}G(x) + \frac{1}{1+x}$ has an asymptotic expansion in $\{x^n\}$ for x small and which has coefficients of $O(1)$ for a small number of terms and of $O(n!)$ for $n >> 100$ terms. Even for asymptotic expansions which occur in a more natural way certain difficulties appear. We will shortly show that

$$G(z) \sim \sum (-1)^i z^i i!$$

for $z \to 0$, $|\arg z| < \frac{3\pi}{2}$ and that the rays $\arg z = \pm \frac{3\pi}{2}$ are Stokes lines for the AD. Although this is a perfectly good AD the estimate of the error becomes poor as the Stokes lines are approached, i.e., using the first neglected term for an estimate becomes less and less accurate as the neighborhood of the Stokes lines is approached. As another aspect of this same type of problem we point out that in expanding in an AS, we are actually agreeing to discard terms that are not included in this gauge. Thus if we consider

$$f(x) = G(\frac{1}{x}) + e^{-ax}, \quad a > 0$$

for x large we obtain

$$f(x) \sim \sum_i \frac{i!(-)^i}{x^i} .$$

However, when it comes to evaluating f by the AD we may be carrying terms which are small with respect to neglected terms. Thus, for example, to compute $f(10)$, the AD suggests that we carry ten terms, whereas if e.g., $a = \frac{1}{10}$ this would then give a totally inaccurate result, we could avoid this problem by writing

$$f = \sum^{N-1} \frac{i!}{x^{\ell}} + R_N$$

and

$$R_N = \max\left(\frac{N!}{x^N}, e^{-ax}\right)$$

R_N is, therefore, a uniform error bound. The importance of uniform error bounds has been greatly stressed in recent years. [See Error Bounds for Asymptotic Expansion - F. Olver in "Asymptotic Solutions of Differential Equations and Their Applications", C. H. Wilcox, Editor.] Although this point is of obvious importance a detailed treatment falls outside the scope of the course and we will not pursue these matters further.

Now having mentioned all the reservations concerning the above remarks (i) - (iv) we state that they serve as a useful guide (as does the warning about the Stokes lines) for AD in general, and they seem to be valid in a large body of cases which occur in practice even though their proof in any specific case may be extremely difficult.

Another piece of folklore concerns the almost unreasonable success of asymptotic expansions even in non-asymptotic regimes. This is illustrated by the AD of $x!$ obtained for $x \gg 1$ and as shown above to be extremely accurate even for $x = 1$. An examination of the AD of Bessel functions and other special functions shows this same effect.

Exercise 8. For $x \gg 1$, find several terms of the AD of

(a) $$\int_x^\infty \frac{dy}{y^2+e^y}$$

(b) $$\int_0^x \frac{e^y dy}{1+y^2+y^3} .$$

Prove your results.

1.4. Summation of Asymptotic Expansions.

We have already pointed out that a given function can have a variety of AD in terms of a variety of AS -- although it has a unique representation in terms of any one AS. The possibility then presents itself of one of these AD being superior than the others from the point of view of possible accuracy and speed of convergence. For example, we can write

$$I = \frac{1}{1+x} \sim \sum_n (-)^n x^n$$

or

$$I \sim (1-x)\sum_n x^{2n}$$

or

$$I \sim (1-x)(1+x^2)\sum_n x^{4n} .$$

$$\vdots$$

Certainly each succeeding AD converges more rapidly for $0 \le x < 1$. However, if we consider

$$II \sim x - \frac{1}{2}x^2 + \frac{1}{3}x^3 +$$

no obvious resummation speeds up the convergence. We now consider a systematic method due to Euler which frequently speeds up the convergence of a series.

Consider the following mapping known as the Euler transformation,

$$w = \frac{x}{1+x}, \quad x = \frac{w}{1-w}.$$

Applying this to I we obtain

$$I \sim 1 - w (= \frac{1}{1+x})$$

and

$$II \sim w + w^2/2 + w^3/3 + \cdots \left[= \left(\frac{x}{1+x}\right) + \frac{1}{2}\left(\frac{x}{1+x}\right)^2 + \frac{1}{3}\left(\frac{x}{1+x}\right)^3 + \cdots\right].$$

In each case the speed of convergence has been improved dramatically, especially for $x \approx 1$, where the previous APS were very slowly convergent. To understand the effectiveness of the Euler transformation, we first note that in the complex plane the Euler transformation is a linear fractional map which maps

$$\text{Re } z > -1/2$$

in the z-plane into the unit circle

$$|w| < 1$$

in the w-plane. Next we recognize that both I and II have convergent APS for $|z| < 1$ and hence each of these are function elements of analytic functions. The APS of I is in fact $1/(1+z)$ and the APS of II will be recognized as $\ln(1+z)$. The former has a pole at $z = -1$ and the latter a branch point at $z = -1$ (and $z = \infty$). Since in both cases no singularity is mapped into the unit circle of the w-plane, the new series, i.e. in w, converges in unit circle of the w-plane. Finally, since $0 \leq x \leq 1$ is mapped into $0 \leq w \leq \frac{1}{2}$ the reason for the rapid convergence of the w-series becomes clear.

Next, we note that in both the above examples the transformed series converge for $x \geq 1$ which was not so for the untransformed case. The reason for this is clear from the previous paragraph, but in order to attach any significance to this we must make a distinction between

$$f(z) \sim \sum a_i z^i$$

and

$$f(z) = \sum a_i z^i$$

when $\sum a_i z^i$ is a function element, i.e., if the infinite series converges. In the second case we know that the transformed series gives the AC of $f(z)$ but in the first case nothing can be said. To see this consider

$$f(z) = e^{-1/z} + \frac{1}{1+z}$$

for $z \to 0$ and $z \in S^\delta$ (recall $S^\delta = S(\frac{\pi}{2} - \delta)$, $\delta > 0$) then

$$f \sim \sum (-z)^i.$$

The Eulered series for $\operatorname{Re} z \geq 1$ gives us no information about the original function. More generally, the AD of a function in the neighborhood of a point tells us nothing about this function in other neighborhoods of definition.

There is another aspect of the Euler transformation which bears on the "best approximation" limitation of the previous section. In certain cases we can ameliorate this condition. We illustrate this through the function $G(x)$. Eulering the integral representation of G directly we obtain,

$$G(w) = \int_0^\infty e^{-t} \frac{dt}{1 + \frac{wt}{1-w}} = (1-w) \int_0^\infty \frac{e^{-t}dt}{1-w(1-t)} .$$

Formally, we have

$$G(w) \sim \sum w^k (1-w) \int_0^\infty e^{-t}(1-t)^k dt.$$

Using the same procedure as used in analyzing $G(x)$ we may demonstrate (for $0 \leq w < 1$) that the error bound is given by the magnitude of the first neglected term. Then

$$G(w) \sim \sum_{k=0} w^k \int_0^\infty e^{-t}(1-t)^k dt - \sum_{k=0} w^{k+1} \int_0^\infty e^{-t}(1-t)^k dt$$

$$= 1 - \sum_{k=1} w^k \int_0^\infty e^{-t} t(1-t)^{k-1} dt$$

$$= 1 - \sum_{k=1}^\infty w^k \sum_{k=0}^{k-1} \int_0^\infty e^{-t} \frac{(k-1)!(-)^j t^{j+1}}{j!(k-1-j)!} dt.$$

So that

$$G(w) \sim 1 - \sum_{k=1} (-)^k k!\, a_k w^k$$

where

$$a_k = \frac{1}{k} \sum_{j=0}^{k-1} \frac{(-)^{j+k}(j+1)}{(k-1-j)!} = - \sum_{p=0}^{k-1} \frac{(-)^p (k-p)}{p!} \; .$$

In particular, we find,

$$G(w) \sim 1 - w + w^2 - 3w^3 + 11w^4 - 53w^5 + 309w^6 \mp \cdots \; .$$

The improvement is clear from this since the coefficients are smaller (also note that w is smaller) than in the corresponding expansion in x. Also, it can be seen that the AD has meaning for $w = \frac{1}{2}$ $(x = 1)$ which was without sense before.

Further note

$$a_k = -\sum_{p=0}^{k-1} \frac{(-)^p}{p!} + \frac{1}{k} \sum_{p=0}^{k-2} \frac{(-)^p}{p!}$$

and as $k \to \infty$

$$a_k \sim -\frac{1}{e} \; .$$

As before for a fixed value of $x \approx 0$ the best approximation is obtained by halting the AD at the minimal term. If x is sufficiently small we can suppose that k is large enough so that $a_k \sim -e^{-1}$. Then the minimal term is given by the largest k such that

$$kw \leq 1$$

or

$$k \leq \frac{1+x}{x}$$

or

$$k = 1 + [\frac{1}{x}]$$

where $[\alpha]$ denotes the integer part of α. Comparing this with

$$G(x) \sim \sum (-1)^j j!\, x^j$$

we see that since $a_k \sim -e^{-1}$, the minimal term is at least $1/e$ smaller than before and that it is necessary to take one additional term in the approximation. [Actually, the minimal term is smaller than our estimate since $\frac{x}{1+x} < x$.] The AD is still not convergent for any value of x as we will see later the origin is a branch point of $G(z)$. On repeating the Euler transformation n-times we may count on reducing the error bound by e^{-n} - but this is offset by the necessity of taking n additional terms.

If we represent the Euler transformation by

$$Tz = \frac{z}{1+z}$$

then one can easily show

<u>Exercise 9</u>.

$$T^n z = \frac{z}{1+nz} .$$

From this we may effect the n-fold repeated application of the Euler transformation directly. For if z is transformed according to

$$w = \frac{z}{1+\alpha z}$$

in

$$f \sim \sum a_n z^n$$

then one may

Exercise 10. Show,

$$f(w) \sim a_o + \sum_{p=1}^{\infty} A_p w^{p+1}$$

where

$$A_p = \sum_{n=o}^{p} a_{n+1} \binom{p}{p-n} \alpha^{p-n} .$$

Two difficulties in the repeated use of the Euler transformation now become apparent. If α is large, the coefficients can become large and secondly a large number of terms may be necessary. As a matter of practical application we can combat these effects by subtracting off the converging portion of the AD before applying the Euler transformation. Thus, for example, to achieve greater accuracy in the calculation of $G(\frac{1}{n})$, we write

$$G(\tfrac{1}{n}) - \sum_{j=o}^{j=n-1} (-)^j j!(\tfrac{1}{n})^j \sim \sum_{j=n} (-)^j j!(\tfrac{1}{n})^j$$

and Euler the right hand side only.

Exercise 11. Suppose

$$f = \sum a_i z^i$$

and it is known that the singularities of $f(z)$ lie in the hatched region. What transformation can be used to obtain a convergent expansion for z belonging to the unhatched portion of the z-plane. (See sketch below)

Although we do not pursue this subject further, we mention that there are a number of generalizations of this subject, sometimes producing very striking increases in the rate of convergence. [See, D. Shanks, "Non-linear transformations of divergent and convergent sequences". J. Math. and Phys. 34, 1-42 (1955).]

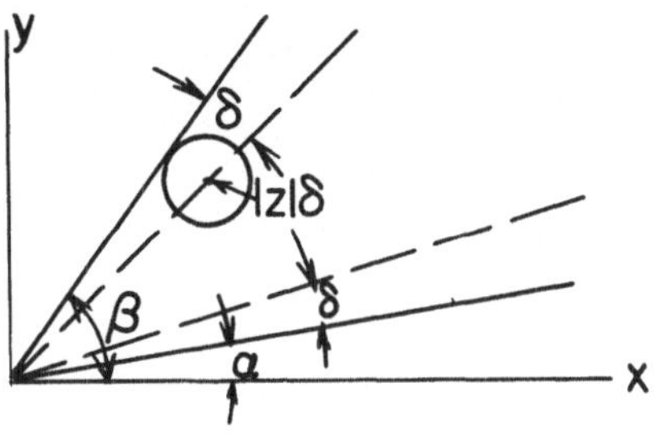

We end this section by briefly considering an associated idea. That of identifying an analytic function with an APS.

A Construction of Borel.

The so-called Borel sum proceeds from a sequence of complex constants $\{a_n\}$, the associated formal series

$$\sum a_i z^{-i}$$

and the identity

$$\int_0^\infty e^{-zt} t^{r-1} dt = \frac{(r-1)!}{z^r}, \quad z \in S^\delta .$$

To see this last formula, first set $zt = \eta$, then

$$\int_0^\infty e^{-zt} t^{r-1} dt = z^{-r} \int_0^{\infty \arg z} e^{-\eta} \eta^{r-1} d\eta$$

and if $z \in S^\delta$ we can bring the path of integration to the real line.

Restricting $z \in S^\delta$ we can write

$$\int_0^\infty e^{-tz} \sum_{i=1}^n \frac{a_i t^{i-1}}{(i-1)!} dt = \sum_{i=1}^n \frac{a_i}{z^i}$$

We now place the restriction on $\{a_i\}$ that

(i) $f(t) = \sum_{i=1}^{\infty} \frac{a_i t^{i-1}}{(i-1)!} < \infty$ for some t_o.

[Then $f(z)$ is defined for $|z| < t_o$] and

(ii) The analytic continuation $f(z)$ is such that

$$S(z) = \int_0^\infty f(t)e^{-tz}dt < \infty.$$

If all of this holds then we obtain a unique analytic function $S(z)$ associated with $\sum a_i z^{-i}$. In the next chapter it is shown that

$$S \sim \sum a_i z^{-i}$$

Actually one may always associate an APS with an analytic function. This is contained in

<u>Theorem 130 (Ritt)</u>. For every APS $\sum a_i z^i$ and sector $S(\overline{S} \supset z = 0)$ one may construct an analytic function $f(z)$ such that

$$f(z) \sim \sum a_i z^i, \quad z \in S.$$

For a proof see Wasow [7]. We note in regard to this theorem that the constructed function is not unique.

Finally using the Borel sum we obtain a criteria for an APS to be the function element of a function analytic in the right half plane.

<u>Theorem 131</u>. $\{b_i\}$ a sequence of real positive constants such that $b_n < M^n$ then there exists an $x_o < 0$ such that

$$\sum_{i=0} (-1)^n b_n z^n$$

defines an analytic function for $\operatorname{Re} z > x_o$.

Proof. Consider the entire function

$$\phi(z) = \sum_{i=o}^{\infty} \frac{(-)^i b_i z^i}{i!} .$$

Further from the hypothesis and well-known estimates we see that there exists an $\varepsilon > 0$ such that

$$\phi = o(e^{-\varepsilon x})$$

for $x \to \infty$. Next consider

$$\hat{f}(z) = \frac{1}{z} \int_o^{\infty} e^{-t/z} \phi(t) dt.$$

This function is analytic for all $z \in S^{\delta}$. Next consider z for $\operatorname{Re} z > 0$ and such that $|z| < \frac{1}{M}$, and hence $|b_n z^n| < 1$, so that

$$\hat{f}(z) = \frac{1}{z} \int_o^{\infty} e^{-t/z} \sum_{i=o}^{\infty} \frac{(-)^i b_i t^i}{i!} dt = \int_o^{\infty} e^{-\eta} \sum_{i=o}^{\infty} \frac{(-)^i (b_i z^i)}{i!} \eta^i d\eta$$

and we may integrate termwise. Finally,

$$\hat{f}(z) = f(z)$$

and $f(z)$ is holomorphic for $\operatorname{Re} z > 0$.

CHAPTER 2.

THE ASYMPTOTIC DEVELOPMENT OF A FUNCTION DEFINED BY AN INTEGRAL.

2.1. Elementary Analytical Methods.

When dealing with an analytic function $f(z)$ it is frequently necessary to know the function for all values of $\arg z$. When the function is defined by an integral this analytic continuation is often facilitated by the following two constructions.

Analytic Continuation of Functions Defined by an Integral

Construction I: Consider

$$F(z) = \int_z^\infty f(w)dw$$

For $|z|$ large, with f analytic and single-valued for $R \leq |z| < \infty$. Then for $|z| \geq R$

$$F(ze^{-i2\pi N}) = F(z) + N \oint_{|w|=R} f(w)dw \tag{2.1.1}$$

where N is an integer and as indicated the contour in the integral is a circle of radius R in the positive direction.

When applicable this construction allows us to restrict attention to the plane covered once. The contour integral (2.1.1) is just a constant, which frequently in practice can be evaluated by elementary mean, e.g. residues.

Example. Consider the function G introduced earlier. By elementary transformations we may write

$$G(\tfrac{1}{z}) = \mathcal{G}(z) = ze^z \int_z^\infty e^{-\zeta} \cdot \frac{d\zeta}{\zeta} \tag{2.1.2}$$

where the path of integration is ultimately along the real line. By Construction I we have

$$(2.1.3) \qquad \mathcal{G}(ze^{-2\pi iN}) = \mathcal{G}(z) + 2\pi iNze^{z}$$

A similar construction is useful in considering the integral

$$H(z) = \int_0^\infty e^{-zw}h(w)dw$$

Although not important we take the path of integration to be the positive axis. Integrals such as these occur naturally as Laplace and Fourier transforms, and frequently other integrals may be reduced to this form.

<u>Construction II:</u> If $h(w)$ is analytic and single valued for $|w| \geq R$ with at most a pole (of arbitrary order) at ∞, then

$$(2.1.4) \qquad H(ze^{-i2\pi}) = H(z) + \oint_{|w|=R} e^{-zw}h(w)dw$$

To see this we first observe that $H(z)$ converges for $z \in S_\delta$. Denote by $P(\phi;R)$ the following path in the w-plane: along the real axis from the origin to $w = R$, along the arc $|w| = R$ from $\arg w = 0$ to $\arg w = \phi$, and from $|w| = R$ to $|w| = \infty$ along the ray $\arg w = \phi$.

Now let $\Delta\phi$ be sufficiently small so that $ze^{-i\Delta\phi} \in S_\delta$ then

$$H(ze^{-i\Delta\phi}) = \int_0^\infty e^{-ze^{-i\Delta\phi}w}h(w)dw = \int_{P(\Delta\phi;R)} e^{-ze^{-i\Delta\phi}w}h(w)dw$$

This follows from

$$\lim_{R\to\infty} \int_{\substack{|w|=R \\ 0 \leq \arg w \leq \Delta\phi}} e^{-ze^{-i\Delta\phi}w}h(w)dw = 0$$

This then provides us with the means by which $H(z)$ may be analytically continued, for by proceeding incrementally we let $\arg z$ increase and ϕ in $P(\phi;R)$ decrease by the same amount so that $\arg(ze^{-i\phi})$ is held fixed. For $ze^{-i\phi} \in S_\delta$ we have

$$H(ze^{-i\phi}) = \int_{P(\phi;R)} \exp[-ze^{-i\phi}w]h(w)dw$$

In particular if z makes a circuit $\phi = 2\pi$ and we obtain (2.1.4) since the integrand is singled valued in z. By repeated use we also have $H(ze^{-2iN\pi}) = H(z) + N \int_{|w|=R} e^{-zw}h(w)dw$.

<u>Note:</u> The only essential way in which e^{-zw} was used in this argument, was as a single valued convergence factor. Therefore the construction also goes through for any function $g(z,w)$ having these properties.

<u>Note:</u> If $f = e^{\gamma w}\hat{f}(w)$, γ a constant, and $\hat{f}(w)$ at most a pole at infinity, the same construction is valid. However as presented above Construction II is valid both for small and large $|z|$.

<u>Integration by Parts</u>

One of the more powerful methods available in asymptotic analysis is simply integration by parts. To start this discussion consider

(2.1.5) $$I(x) = \int_\alpha^\beta F(y)G(xy)dy$$

over the interval (α,β) and for $x >> 1$. Then define

(2.1.6)
$$F_0 = F,\quad F_1 = \frac{d}{dy}F_0, \dots, F_n = \frac{d}{dy}F_{n-1}$$
$$G_0 = G(s),\quad G_{-1}(s) = \int G_0(s)ds, \dots, G_{-n} = \int G_{-n+1}(s)ds$$

Proceeding formally

$$I(x) = \int_{\alpha}^{\beta} F(y) \frac{1}{x} \frac{d}{dy} G_{-1}(xy)dy$$

$$= \frac{1}{x} [F(\beta)G_{-1}(x\beta) - F(\alpha)G_{-1}(\alpha x)] + R_1$$

$$R_1 = -\frac{1}{x} \int_{\alpha}^{\beta} F_1(y)G_{-1}(xy)dy$$

and in general

(2.1.7)
$$I(x) = \sum_{i=1}^{N} (-1)^{i-1} x^{-i} [F_{i-1}(\beta)G_{-i}(x\beta) - F_{i-1}(\alpha)G_{-i}(x\alpha)] + R_N = S_N + R_N$$

with

(2.1.8)
$$R_N = (-1)^N x^{-N} \int_{\alpha}^{\beta} F_N(y)G_{-N}(xy)dy$$

From the formal construction we have

<u>Theorem 210.</u> For (α,β) a finite integral, G bounded say $|G| < M$, and F K-times differentiable we have

(2.1.9)
$$I = S_N + O(x^{-N-1})$$

for all $N < K$.

<u>Proof.</u> Since $|G| < M$ then clearly $|G_{-1}| \le (\beta-\alpha)M$ and $|G_{-k}| \le (\beta-\alpha)^k M$ so that all indefinite integrals of G are bounded. Then writing

$$I = S_N + S_{N+1} - S_n + R_{N+1}$$

we easily see that

$$S_{N+1} - S_N = O(x^{-N-1})$$

$$R_N = O(x^{-N-1})$$

which proves (2.1.9).

As an application consider

$$(2.1.10) \qquad I(x) = \int_0^1 e^{ixy} F(y)dy$$

In this case

$$G(s) = e^{is}$$

and

$$G_{-n}(s) = (-i)^n e^{is}$$

For sufficiently differentiable $F(y)$ all the hypothesis of the theorem hold and

$$(2.1.11) \qquad I(x) \sim - \sum_{n=1} i^n x^{-n} [F_{n-1}(1)e^{ix} - F_{n-1}(0)]$$

As an immediate extension consider

$$(2.1.12) \qquad \int_a^b e^{ixh(t)} g(t)dt$$

where $h(t)$ is real and $h'(t) \neq 0 \quad t \in [\alpha,\beta]$. Then setting

$$y = h(t)$$

$$\beta = h(b)$$

$$\alpha = h(a)$$

we obtain

(2.1.13) $$\int_{\alpha}^{\beta} e^{ixy} \frac{g(t(y))}{h'(t(y))} dy$$

and above application can again be used.

Exercise 12. Consider $I(x) = \int_{-\infty}^{+\infty} e^{ixy} g(y)dy$

(a) $g \in C_0^\infty$ (infinitely differentiable and compact support, i.e. identically zero outside some finite portion of the real line) then

$$I \sim 0 \text{ to all orders.}$$

(b) $g \in C_0^\infty$ except for a finite discontinuity at the origin

$$I \sim O(\tfrac{1}{x}).$$

(c) $g \in C_0^\infty$ except at the origin where $g \sim \frac{1}{|y|^r}$, $r < 1$ then

$$I \sim \frac{1}{x^{-r+1}}$$

The cases, $h' = 0$ in the interval of integration and infinite intervals of integration have to be treated more carefully. These we discuss later.

For the case just treated the way in which parts integration should enter is clear from the location of the large parameter. In general however the issue becomes subtler and a more delicate use of parts integration is required. Of use are the following two identities;

(2.1.14) $$g = \frac{d}{dt}(tg(t)) - tg'(t)$$

(2.1.15) $$g = g'h = \frac{d}{dt}(gh) - gh'$$

The first equality in the second relation defines h. Relations (2.1.14) and (2.1.15) are in essence the only relations used in the following discussion.

Asymptotic Evaluation of Indefinite Integrals $\int^x f(t)dt$ (The following is based on a similar discussion given in Dieudonne [13])

Recalling the discussion leading up to Theorem 126, we restrict attention to integrals along the real line of types

$$\int_x^\infty f(t)dt \quad \text{and} \quad \int_a^x f(t)dt$$

with f a complex valued function. Under the conditions of Theorem 126 parts A3 and B3 we can consider integrals of the form $\int g(t)dt$ if $f \sim cg$. However this still leaves us with no guarantee that $\int g dt$ can be exactly integrated and in general it cannot be, even for the important class (see (1.2.1))

$$f \sim cx^{\alpha}(\ell n\ x)^{\beta}e^{P(x)} = cg, \quad x \to \infty \tag{2.1.16}$$

mentioned earlier. The purpose of the calculations given below is to demonstrate that we can systematically give an AD for integrals $\int f dt$ which includes functions of the class (2.1.16). Specifically we consider the admissibility class: $g > 0$, differentiable and

$$\frac{g'}{g} \sim \frac{\beta}{x\,\ell n\ x} + \frac{\alpha}{x} + P'(x) \tag{2.1.17}$$

where

$$P(x) = \sum_{i=1}^{k} c_i x^{\gamma}$$

with $\gamma_1 > \gamma_2 > \dots \gamma_k > 0$ and α, β, c_i real.

Case I: g admissible and

$$\frac{g'}{g} \sim \frac{\mu}{x} \tag{2.1.18}$$

with $\mu \neq 0, -1$.

<u>I.a:</u> $\mu > -1$, then $x^{\mu-\varepsilon} = O(g)$ for all $\varepsilon > 0$, $\varepsilon \to 0$ as $x \to \infty$, and

$$\int_a^x g(t)dt \sim \frac{xg(x)}{\mu+1} \tag{2.1.19}$$

<u>I.b:</u> $\mu < -1$, then $g = O(x^{\mu+\varepsilon})$ for all $\varepsilon > 0$ as $x \to \infty$ and

$$\int_x^\infty g(t)dt \sim -\frac{xg(x)}{\mu+1} \tag{2.1.20}$$

<u>Proof I.a:</u> From Theorem <u>126</u> <u>A3</u>

$$\ln g \sim \mu \ln x$$

Therefore for any $\varepsilon > 0$ and x sufficiently large

$$\ln g \geq (\mu - \frac{\varepsilon}{2}) \ln x$$

and exponentiating we have

$$g \geq x^{\mu - \frac{\varepsilon}{2}}$$

for any $\varepsilon \to 0$. In particular taking ε sufficiently small demonstrates that (2.1.19) is the correct indefinite integral to consider.

Next integrating (2.1.14) we obtain,

$$\int_a^x g(t)dt = xg(x) - ag(a) - \int_a^x tg'(t)dt$$

and rewriting this

$$\int_a^x [g(t) + tg'(t)]dt = xg(x) - ag(a)$$

and from (2.1.18) $tg' \sim \mu g$. Therefore from Theorem 126 A3

$$\int_a^x [g(t) + tg'(t)]dt \sim (\mu+1)\int_a^x g(t)dt$$

and (2.1.19) follows.

Proof (I.b): Again we have

$$\ln g \sim \mu \ln x$$

and therefore for any $\varepsilon > 0$ and x sufficiently large

$$\ln g \leq (\mu+\varepsilon) \ln x$$

and on exponentiating

$$g \leq x^{\mu+\varepsilon}$$

Choosing $\varepsilon > 0$ sufficient small we see that (2.1.20) is the proper indefinite integral to consider.

Again integrating (2.1.14)

$$\int_x^\infty g(t)dt = -xg(x) - \int_x^\infty tg'(t)dt$$

and applying the analogous argument (2.1.20) follows.

To take care of the case of when $\mu = 0$ in (2.1.18) we have,

Corollary. Suppose

(2.1.21) $$\frac{g'(x)}{g(x)} = o(\frac{1}{x})$$

then by Theorem 126 A2,

$$\ln g = o(\ln x)$$

From this it follows that for any $\varepsilon > 0$

$$|\ln g| < \varepsilon \ln x$$

for x sufficiently large.

This implies

$$\varepsilon \ln x \geq -\ln g$$

and therefore

$$g > x^{-\varepsilon}$$

for all $\varepsilon > 0$, and we should consider

$$\int_a^x g(t)dt$$

Integrating (2.1.14)

$$\int_a^x (g(t) + tg'(t))dt = xg(x) - ag(a)$$

But in view of the hypothesis

$$tg'(t) = o(g)$$

and then again from Theorem 126 A3

$$(2.1.22) \qquad \int_a^x g(t)dt \sim xg(x),$$

or in other words (2.1.19) still applies.

Example. Consider

$$\int_a^x \frac{dt}{\ln t}$$

then

$$\frac{g'(t)}{g} = -\frac{1}{t \ln t}$$

and the above discussion applies and therefore

$$\int_a^x \frac{dt}{\ln t} \sim \frac{x}{\ln x} .$$

Example. Consider

$$g(t) = t^2 \exp\left(\frac{t}{1+t}\right)^{\frac{1}{2}}$$

then

$$\frac{g'}{g} \sim \frac{2}{t}$$

and Case Ia applies and we have

$$\int_a^x g(t)dt \sim \frac{x^3 e^{\left(\frac{x}{1+x}\right)^{\frac{1}{2}}}}{3}$$

We now take up the other situation, namely

$$\frac{1}{x} = o\left(\frac{g'}{g}\right).$$

Also we will assume that $g' \neq 0$ for x sufficient large. (This is certainly in keeping with g of the type (2.1.16).) From this we have that

$$h = \frac{g}{g'}$$

is defined for x large.

Case II. Suppose $h = g/g'$ is differentiable at ∞ and also that

$$h'(x) = o(1) \tag{2.1.23}$$

for $x \to \infty$.

II.a: If $g'(x) > 0$ in the neighborhood of ∞, then $\int^{\infty} g dt = \infty$ and

$$\int_a^x g(t)dt \sim \frac{(g(x))^2}{g'(x)} \tag{2.1.24}$$

II.b: If $g'(x) < 0$ in the neighborhood of ∞, $\int^{\infty} g(t)dt < \infty$ and

$$\int_x^{\infty} g(t)dt \sim -\frac{(g(x))^2}{g'(x)} \tag{2.1.25}$$

To begin we note that from Theorem 126 A2, (2.1.23) implies $h = o(x)$. Now in Case IIa, $g' > 0$, and this may be written as

$$\frac{1}{x} = o\left(\frac{g'}{g}\right)$$

and then again applying Theorem 126

$$\ln x = o(\ln g)$$

or

$$\ln x < \varepsilon \ln g$$

for all $\varepsilon > 0$ and

$$g > x^{\alpha}$$

for any $\alpha > 0$. In Case IIb, $g' < 0$

$$\ln x < \varepsilon \, |\ln g|$$

In this case $\ln g < 0$ otherwise a contradiction and

$$-\frac{1}{\varepsilon} \ln x > \ln g$$

$$g < x^{-\alpha}$$

for all $\alpha > 0$. Therefore in Case IIa, $\int^{\infty} g(t)dt = \infty$, and in Case IIb, $\int^{\infty} g(t)dt < \infty$.

<u>Proof of IIa:</u> Integrating (2.1.15)

$$\int_a^x g(t)dt = \int_a^x h(t)g'(t)dt = h(x)g(x) - h(a)g(a) - \int_a^x h'(t)g(t)dt$$

and therefore

$$\int_a^x (1 + h'(t))g(t)dt = h(x)g(x) - h(a)g(a)$$

But from (2.1.23)

(2.1.26) $$(1 + h')g \sim g$$

Then applying theorem 126 A3, we obtain (2.1.24).

Proof of IIb: Again integrating (2.1.15)

$$\int_x^\infty [1 + h'(t)]g(t)dt = -h(x)g(x)$$

Then from (2.1.26) and Theorem 126.B3 (2.1.25) follows.

So far we have excluded the case

$$\frac{g'}{g} = -\frac{1}{x} + o(\frac{1}{x}) \tag{2.1.27}$$

Actually this case is not difficult although it can be tedious. For example if g is of the class (2.1.16) then (2.1.27) states that

$$g \sim \frac{(\ln x)^\beta}{x}$$

and in this case we make use of the well-known integrals,

$$\int \frac{(\ln t)^\beta}{t} dt = \frac{(\ln x)^{\beta+1}}{\beta+1}, \quad \beta \neq -1 \tag{2.1.28}$$

$$\int \frac{dt}{t \ln t} = \ln \ln x \tag{2.1.29}$$

If g is not of the class (2.1.16) a transformation of the variable of integration should be attempted, e.g. consider

$$g = \frac{1}{t \ln \ln t}$$

for which

$$\frac{g'}{g} \sim -\frac{1}{t}$$

considering

$$\int_a^x \frac{dt}{t \ln \ln t}$$

we set $\ln t = s$ and find

$$\int_{\ln a}^{\ln x} \frac{ds}{\ln s}$$

and from an example given above.

$$\int_a^x \frac{dt}{t \ln \ln t} \sim \frac{\ln x}{\ln \ln x}$$

Although the above discussion focused on finding the leading term in the asymptotic development, it also can be used to obtain subsequent terms in the AD. For example consider g such that

$$\frac{g'}{g} \sim \frac{\mu}{x}, \quad \mu > -1$$

and hence the first step is

$$\int_a^x g(1 + \frac{tg'}{g})dt = xg(x) - ag(a)$$

or

$$\int_a^x g(1+\mu)dt + \int_a^x g(\frac{tg'}{g} - \mu)dt = xg(x) - ag(a).$$

Writing

(2.1.30) $$g_1 = \frac{tg'(t)-\mu g(t)}{\mu+1}$$

we obtain

$$\int_a^x g\,dt = \frac{xg(x)-ag(a)}{\mu+1} - \int_a^x g_1(t)dt \tag{2.1.31}$$

The entire procedure now applied to the integral $\int g_1(t)dt$. In particular if $\int^{\infty} g_1 dt < \infty$ we write

$$\int_a^x g_1(t)dt = \int_a^{\infty} g_1(t)dt - \int_x^{\infty} g_1(t)dt.$$

At each step we are led to a new problem of the type already considered.

Illustration: We illustrate this procedure with the following example. Consider

$$g(t) = t^{-1/2}\tan^{-1}t, \quad G(x) = \int_0^x g(t)dt$$

(we take the principal branch of $\tan^{-1}t$, i.e. $\lim_{t\to\infty} \tan^{-1}t = \frac{\pi}{2}$). Then

$$g' = -\frac{\tan^{-1}t}{2t^{3/2}} + \frac{1}{t^{1/2}(1+t^2)}$$

so then

$$\frac{g'}{g} \sim -\frac{1}{2t}, \quad \mu = -\frac{1}{2}$$

Next we note that

$$\begin{aligned} g_1 &= 2\{-\frac{\tan^{-1}t}{2t^{1/2}} + \frac{t^{1/2}}{1+t^2} + \frac{\tan^{-1}t}{2t^{1/2}}\} \\ &= \frac{2t^{1/2}}{1+t^2}. \end{aligned}$$

Therefore

$$G(x) = 2x^{1/2}\tan^{-1}x - \int_0^x g_1(t)dt.$$

Now since $\int^{\infty} g_1(t)dt < \infty$, we rewrite as

$$G(x) = 2x^{1/2}\tan^{-1}x - 2\int_0^{\infty} \frac{t^{1/2}}{1+t^2}dt + 2\int_x^{\infty} \frac{t^{1/2}dt}{1+t^2}$$

$$= 2x^{1/2}\tan^{-1}x - \pi\sqrt{2} + 2\int_x^{\infty} \frac{t^{1/2}dt}{1+t^2} .$$

The remaining integral may be exactly integrated also we can obtain the asymptotic development from

$$\int_x^{\infty} \frac{t^{1/2}}{1+t^2} dt \sim \int_x^{\infty} \frac{1}{t^{3/2}} \sum_{k=o} (-1)^k t^{-2k} dt$$

In dealing with more difficult problems the rule is that at each successive step we apply the general procedure to the remainder function g_1. At each such step we must test to see whether $\int^{\infty} g_1(t)dt$ converges - and then use the appropriate indefinite integral. The same procedure also applies to indefinite integrals falling under Case II above. In this case we have, e.g.

$$\int_a^x g(t)dt = h(x)g(x) - h(a)g(a) - \int_a^x g_1(t)dt$$

with $g_1 = g(t)h'(t)$. A point of note is that because of Theorem <u>126</u> <u>A,B3</u> we can always deal with the lead term in the AD of g_1 if this proves more convenient.

<u>Asymptotic Evaluation of Integrals of the Form $\int^x f(x,t)dt$.</u>

The above methods and discussion given in Cases I and II often apply to integrals of the type

(2.1.32)
$$\int_x^{\infty} g(t,x)dt, \quad \int_a^x g(t,x)dt.$$

in which the large quantity $x \to \infty$ appears also in the integrand. The situation is greatly complicated by the fact that both t and x approach infinity. Although certain general results may be obtained (see the exercises) we explore these cases by example, using the above Cases I and II as a guide.

Consider

$$F(x) = \int_0^x (t^3+t^2+x^2)^{1/2}dt.$$

Formally this falls under Case I, part I.a with $\mu = \frac{3}{2}$. We therefore consider (2.1.31)

$$\int_0^x g(t,x)dx = \frac{xg(x,x)}{\mu+1} - \int_0^x \frac{tg_t(t,x)-\mu g(t,x)}{\mu+1}dt$$

and must show that the last term is little o of the first term on the right hand side. In particular we write,

$$\int_0^x (t^3+t^2+x^2)^{1/2}dt = \frac{2x(x^3+2x^2)^{\frac{1}{2}}}{5} - \int_0^x g_1(t)dt$$

with

$$g_1 = \frac{2}{5}\left[\frac{1}{2}\frac{3t^3+2t^2}{(t^3+t^2+x^2)^{\frac{1}{2}}} - \frac{3}{2}(t^3+t^2+x^2)^{\frac{1}{2}}\right]$$

$$= \frac{-t^2-3x^2}{5(t^3+t^2+x^2)^{\frac{1}{2}}}$$

But

$$\int_0^x \frac{t^2+3x^2}{(t^3+t^2+x^2)^{\frac{1}{2}}}dt < \int_0^x \frac{t^2+3x^2}{(t^3+t^2)^{\frac{1}{2}}}dt = O(x^{3/2})$$

and hence

$$\int_0^x (t^3+t^2+x^2)^{1/2}dt \sim \frac{2x^{5/2}}{5}$$

As an example which falls under Case II, consider

$$g(t,x) = \exp\left[\frac{(t^3+x)^{\frac{1}{2}}}{1+x}\right], \quad G(x) = \int_1^x g(t,x)dt.$$

Formally this falls under Case IIa. Parts integrating in the indicated way we write

$$\int_1^x g(t,x)dt = \frac{[g(x,x)]^2}{\frac{\partial}{\partial t} g(t,x)\Big|_{t=x}} - \frac{[g(1,x)]^2}{\frac{\partial g}{\partial t}(t,x)\Big|_{t=1}} - \int_1^x g(t,x)\left[\frac{g(t,x)}{g_t(t,x)}\right]_{,t} dt$$

and we must show that the last integral is little o of the first of the right hand side. Calculating the various quantities

$$g_t = \frac{3t^2 e^{\frac{(t^3+x)^{\frac{1}{2}}}{(1+x)}}}{2(1+x)(t^3+x)^{1/2}}$$

$$\frac{g}{g_t} = \frac{2(1+x)(t^3+x)^{1/2}}{3t^2}$$

$$\frac{\partial}{\partial t}\left(\frac{g}{g_t}\right) = \frac{(1+x)}{(t^3+x)^{1/2}} - \frac{4(1+x)(t^3+x)}{3t^3(t^3+x)^{1/2}}$$

We therefore have

$$\frac{[g(x,x)]^2}{g_t(x,x)} \sim \frac{2\sqrt{x}\, e^{\sqrt{x}}}{3}$$

$$\frac{[g(1,x)]^2}{g_t(x,x)} \sim \frac{2x^{2/3}}{3}$$

We must now demonstrate that the terms in the remainder integral are small. For example there occurs

$$\int_1^x \frac{x^2 e^{\frac{(t^3+x)^{\frac{1}{2}}}{1+x}} dt}{t^3(t^3+x)^{1/2}} \leq \int_1^x \frac{dt}{t^{9/2}}\, e^{\frac{(x^3+x)^{\frac{1}{2}}}{1+x}}\, x^2 = O\left(\frac{e^{\sqrt{x}}}{x^{3/2}}\right)$$

There is however one troublesome term

$$x\int_1^x \frac{e^{\frac{(t^3+x)^{\frac{1}{2}}}{1+x}}}{(t^3+x)^{\frac{1}{2}}}\,dt$$

which arises from both terms in the above expression for $\frac{\partial}{\partial t}[\frac{g}{g_t}]$. To estimate this we write

$$x\int_1^x \frac{e^{\frac{(t^3+x)^{\frac{1}{2}}}{1+x}}}{(t^3+x)^{\frac{1}{2}}}\,dt = x\int_1^{x/2} \frac{e^{\frac{(t^3+x)^{\frac{1}{2}}}{1+x}}}{(t^3+x)^{\frac{1}{2}}}\,dt + x\int_{\frac{x}{2}}^x \frac{(t^3+x)^{\frac{1}{2}}}{(t^3+x)^{\frac{1}{2}}}\,dt$$

Consider the first integral of the right hand side. Replacing t by 1 in denominator and by $x/2$ in the numerator

$$x\int_1^{x/2} \frac{e^{\frac{(t^3+x)^{\frac{1}{2}}}{1+x}}}{(t^3+x)^{\frac{1}{2}}}\,dt \le \sqrt{x}\, e^{\frac{\sqrt{x}}{2^{3/2}}}$$

and hence is negligible. In second integral replace t by $(\frac{x}{2})$ and therefore

$$\int_{\frac{x}{2}}^{x} \frac{e^{\frac{(t^3+x)^{\frac{1}{2}}}{1+x}}}{(t^3+x)^{\frac{1}{2}}}\,dt \le O(\frac{1}{\sqrt{x}}\, G(x))$$

This therefore demonstrates that

$$G(x) \sim \frac{2\sqrt{x}\, e^{\sqrt{x}}}{3}$$

Examples involving Cases Ib and IIb are given in the exercises. In all instances it is a question of being able to show that the formal remainder integral is small compared to the leading term of the parts integration.

To complete this section we will consider an integral in the complex plane. A number of results in Theorem 126 no longer apply and we have to proceed with caution. As a basic difficulty we mention that although integrals of the form

$$\int_z^{\infty} f(\zeta)d\zeta$$

usually end up along straight lines, the path needed to reach this straight line may be $O(|z|)$ in length. This usually ruins many estimates used in Theorem 126. Nevertheless the procedures of Cases I and II can be followed with profit - it is now important however to examine the error terms more carefully.

We illustrate these points by completing our discussion of the function G, (1.1.3). In particular we consider the form (2.1.2)

$$\mathcal{G}(z) = ze^{-z}\int_z^{\infty} e^{-\zeta}\,\frac{d\zeta}{\zeta}$$

It is clear that this falls under IIb.

As has been demonstrated at the outset of this section the origin is a branch point of $\mathcal{G}(z)$, and it is convenient to place a branch cut along the negative axis to $-\infty$, and by virtue of Construction I, p. 38, it suffices to restrict attention to $|\arg z| < \pi$. As for the path of integration any which ultimately goes to infinity along the positive axis is satisfactory, but it will be useful to choose the specific path Γ: arc $|\zeta| = |z|$ from $\arg \zeta = \arg z$ to $\arg \zeta = 0$ and $\zeta = |z|$ to $\zeta = \infty$.

Instead of following Case IIb we integrate the integrand of $\mathcal{G}(z)$ directly,

$$\begin{aligned}\mathcal{G}(z) &= ze^{z}\int_z^{\infty}\left(-\frac{1}{\zeta}\right)\frac{d}{d\zeta}(e^{-\zeta})d\zeta \\ &= e^{z}\,z\left[\frac{e^{-z}}{z} - \int_z^{\infty}\frac{e^{-\zeta}}{\zeta^2}d\zeta\right]\end{aligned}$$

And continuing

$$\mathcal{G}(z) = \sum_{i=0}^{k}\frac{(-1)^i i!}{z^i} + R_k = S_k + R_k \qquad \text{with}$$

(2.1.33)

$$R_k = (-1)^{k+1}(k+1)!\ z \int_z^\infty \frac{e^{z-\zeta}}{\zeta^{k+2}}\, d\zeta$$

Of course the expression for S_k has been obtained already for $z = x > 0$ and in that case we showed $|R_k| < (k+1)!x^{-(k+1)}$. In the present case, assuming $\arg z > 0$

$$|R_k(z)| = (k+1)!|z| \left| \int_{\substack{\zeta = |z| \\ \arg z \geq \arg \zeta \geq 0}} e^{z-\zeta} \frac{d\zeta}{\zeta^{k+2}} + \int_{|z|}^{\infty} e^{z-\zeta} \frac{d\zeta}{\zeta^{k+2}} \right|$$

$$|R_k(z)| \leq (k+1)!|z| \{ \int_0^{\arg z} |e^{z-\zeta}| \frac{|d\zeta|}{|\zeta^{k+2}|} + \int_{|z|}^{\infty} e^{|z|-\zeta} \frac{d\zeta}{\zeta^{k+2}} \}$$

In regard to the first integral note

$$|e^{z-\zeta}| = e^{x-\mathrm{Re}\,\zeta} \leq 1$$

and $|\zeta| = |z|$, and in the second integral it will be conservative to take $|\zeta| = |z|$. Therefore

$$|R_k(z)| \leq \frac{(k+1)!}{|z|^k} (\pi + \frac{1}{|z|})$$

It should be clear that it is the path length of $O(|z|)$ in the first integral which ruins the estimate. Now a "bootstrap" argument is useful, since

$$R_k = (-1)^{k+1}(k+1)!\,z^{-(k+1)} + R_{k+1}$$

and hence

$$(2.1.34) \qquad |R_k| \leq \frac{(k+1)!}{|z|^{k+1}} [1 + (k+2)\pi + \frac{k+2}{|z|}]$$

Although we have demonstrated that $\Sigma(-1)^i i!\,z^{-i}$ is the AD of $\mathscr{G}(z)$ our error estimate is not as sharp as before. In particular to say that the error is given by the magnitude of the first neglected term may no longer be valid.

Next consider the function $\mathscr{G}(z)$ for $\pi \leq \arg z \leq 2\pi$. By (2.1.3) we have

$$\mathscr{G}(z) = \mathscr{G}(ze^{-2\pi i}) - 2\pi i z e^{z}$$

But $-\pi \leq \arg ze^{-2\pi i} \leq 0$ and we can use (2.1.33-34) in this range. Therefore

$$\mathscr{G}(z) = \sum_{i=0}^{k} \frac{(-1)^i i!}{z^i} + R_k(z) - 2\pi i z e^{z} \qquad \text{for } \pi \leq \arg z \leq 2\pi \tag{2.1.35}$$

But e^z is asymptotically small in the third quadrant, and on repeating the above for $-2\pi < \arg z \leq -\pi$, we obtain

$$\mathscr{G}(z) \sim \sum_{i=0}^{k} \frac{(-1)^i i!}{z^i} \qquad |\arg z| \leq \frac{3\pi}{2} - \delta, \quad \delta > 0 \tag{2.1.36}$$

The line $\arg z = \frac{3\pi}{2}$ is clearly a Stokes line, and

$$\mathscr{G}(z) \sim -2\pi i z e^{z}$$

$$\frac{3\pi}{2} < \arg z < \frac{5\pi}{2}$$

This extended sector follows from continuation, considering $\mathscr{G}(z)$ for $2\pi < \arg z \leq 3\pi$. By continuing this discussion the analysis of $\mathscr{G}(z)$ for all $\arg z$ may be given. One other point of note is that although (2.1.36) tells us that the AD is valid in $\pi \leq \arg z < \frac{3\pi}{2}$, the error now is

$$|R_k - 2\pi i z e^{z}|$$

which becomes increasingly large as $\frac{3\pi}{2}$ is approached. This is a concrete example of the remarks given on page 26.

Exercise 13. Obtain the leading term in the expansion at ∞ of the following integral

$$\int_0^x \frac{e^{-1/t^{3/2}}}{(1+t)^{\frac{1}{2}}} dt$$

Exercise 14. Find two terms in the AD for $x \to \infty$ of

$$\int_1^x e^{-1/(\ln t)^{\frac{1}{2}}} dt$$

Exericse 15. Find

$$\int_2^x \frac{dt}{\ln t}$$

for $x \to \infty$ to all orders.

Exercise 16. Find three orders in the AD of

$$\int_a^x \frac{e^t}{t^2+1} dt$$

as $x \to \infty$.

Exercise 17. Suppose $g(t,x) > 0$ is such that for x fixed

$$\frac{g_t(t,x)}{g(t,x)} \sim \frac{\mu}{t}, \quad \mu < -1$$

independently of x. Prove

$$\int_x^\infty g(t,x)dx \sim -\frac{xg(x,x)}{\mu+1}$$

Exercise 18. Under the same hypothesis as the above, but with $\mu > -1$ show that

$$\int_a^x g(t,x) \sim \frac{xg(x,x)}{\mu+1}$$

can be false. Can you find a general rule?

Exercise 19. Find the leading term in the AD of

$$\int_1^x e^{t^t} dt.$$

Exercise 20. Find AD of

$$E_n(z) = \int_1^\infty \frac{e^{-zt}}{t^n} dt$$

for all arg z.

Exercise 21. Find AD of

$$S_i(z) = \int_0^z \frac{\sin t}{t} dt$$

for $|z|$ large and small.

2.2. Laplace and Fourier Transforms at Infinity

Definition. The following integral transform of $\psi(t)$

$$\Phi(z) = \mathcal{L}(\psi) = \int_0^\infty e^{-zt}\psi(t)dt \tag{2.2.1}$$

is referred to as the Laplace transform of $\psi(t)$.

Definition. If $\mathcal{L}(\psi)$ exists for $z = z_0$ then ψ is said to be of class $\mathcal{L}_{z_0}$, written $\psi \in \mathcal{L}_{z_0}$.

Theorem 220. If $\psi \in \mathcal{L}_{z_0}$ then

(a) $\psi \in \mathcal{L}_z$ for all z such that $\operatorname{Re} z > \operatorname{Re} z_0$.

(b) $\mathcal{L}(\psi)$ is analytic for $\operatorname{Re} z > \operatorname{Re} z_0$.

Proof. Set $f(z_0;T) = \int_0^T e^{-z_0 s}\psi(s)ds$, $T > 0$. Then since $\psi \in \mathcal{L}_{z_0}$, $f(z_0;T)$ is bounded for all T. Consider

$$\begin{aligned}\phi(z) &= \lim_{T\to\infty}\int_0^T e^{-(z-z_0)t}\psi(t)e^{-z_0 t}dt \\ &= \lim_{T\to\infty}\int_0^T e^{-(z-z_0)t}\frac{d}{dt}\int_0^t \psi(s)e^{-z_0 s}ds\,dt \\ &= \lim_{T\to\infty}\int_0^T (z-z_0)e^{-(z-z_0)t}f(z_0;t)dt.\end{aligned}$$

From the boundedness of $f(z_0;t)$ it follows that the limit exists for $\operatorname{Re}(z-z_0)>0$, which gives us part a.

Next we observe that for any bounded set in $\operatorname{Re}(z-z_0) > 0$, the convergence of this is uniform. Next we note that

$$\int_0^T e^{-zt}\psi(t)dt$$

is analytic for any $T < \infty$. Part b follows now since the limit of a uniformly convergent sequence of analytic functions is itself analytic.

In this section we will consider (2.2.1) for $|z|$ large. As a first result in this direction we quote

Theorem 221. (Riemann-Lebesgue): For $\psi(t)$ such that

$$\int_0^\infty |\psi|dt < \infty \tag{2.2.2}$$

and y real, then

$$\int_{-\infty}^\infty e^{-iyt}\psi(t)dt = o(1)$$

as $|y| \to \infty$.

(See e.g. [6] for a proof.)

The following almost trivial extension is useful.

Theorem 222. (Generalized Riemann-Lebesgue): For ψ satisfying (2.2.2),

$$\phi = \mathcal{L}(\psi) = o(1)$$

when $\operatorname{Re} z \geq 0$ and $|z| \to \infty$.

To prove this we write

$$|\phi| = \left|\int_0^a \psi(t)e^{-zt}dt + \int_a^\infty \psi(t)e^{-zt}dt\right|$$

where $a > 0$ is not yet specified. Then setting $z = x+iy$,

$$|\phi| \leq \int_0^a |\psi| dt + e^{-xa} \int_a^\infty |\psi| dt.$$

Therefore choosing $a > 0$ sufficient small and x_ε sufficiently large we have

$$|\phi| < \varepsilon$$

for $x > x_\varepsilon$.

Next we observe that ψe^{-xt} for $x \geq 0$ is also absolutely integrable and hence by Riemann- Lebesgue, Theorem 221, there exists a $y(\varepsilon,x)$ such that

$$|\phi| < \varepsilon$$

for $|y| > y(\varepsilon,x)$. Letting $y_\varepsilon = \max_{0 \leq x \leq x_\varepsilon} y(\varepsilon,x)$ we have demonstrated

$$|\phi| < \varepsilon \text{ if either } x > x_\varepsilon \text{ or } y > y_\varepsilon,\ x_\varepsilon \geq x \geq 0$$

which proves Generalized Riemann-Lebesgue.

Corollary. If $\psi = e^{bt}\,\chi(t)$ and χ satisfies (2.2.2) then $\mathcal{L}(\psi) = o(1)$ for $|z| \to \infty$ and $\mathrm{Re}(z-b) \geq 0$.

Corollary. If ψ satisfies (2.2.2) and $\psi_a = H(t-a)\psi$, H is the Heaviside operator, i.e., $H(t) = 0$ if $t < 0$ and $=1$ otherwise) then

$$\phi_a = \mathcal{L}(\psi_a) = o(\exp az)$$

for $|z| \to \infty$ and $\mathrm{Re}\, z \geq 0$.

Proof. $e^{az}\phi_a = \int_0^\infty e^{-zt}\psi(t+a)dt.$

Watson's Lemma

Watson's lemma which we obtain now gives us specific information on the behavior of $\mathcal{L}(\psi)$ at ∞.

Watson's Lemma (1st form). For ψ satisfying (2.2.2) and $\psi = o(t^\gamma)$, $\mathrm{Re}\,\gamma > -1$ as $t \to 0$, then

$$\mathcal{L}(\psi) = o(z^{-(\gamma+1)})$$

for $|z| \to \infty$ and $z \in S^\delta$.

Proof. We wish to demonstrate that for $z \in S^\delta$, $|z| \to \infty$ and arbitrary $\varepsilon > 0$

$$|\mathcal{L}(\psi)| \leq \varepsilon |z^{-(\gamma+1)}|$$

On writing $z = |z|\,e^{i\theta} = e^{\ell n|z| + i\theta}$ and $\gamma = \alpha + i\beta$

$$|z^{-(\gamma+1)}| = |e^{-(\ell n\, z + i\theta)(\alpha+i\beta+1)}|$$

$$= |z|^{-(\alpha+1)}e^{+\beta\theta}$$

the above condition for proof becomes

$$|\mathcal{L}(\psi)| \leq \varepsilon |z|^{-(\alpha+1)} e^{+\beta\theta}$$

so that we equivalently want to show

$$\mathcal{L}(\psi) = o(z^{-(\alpha+1)})$$

Choosing $a > 0$ sufficiently small so that

$$|\psi(t)| < \varepsilon |t^{\gamma}| = \varepsilon t^{\alpha}, \ 0 \leq t \leq a$$

we have

$$|\mathcal{L}(\psi)| \leq \varepsilon \int_0^a e^{-xt} t^{\alpha} dt + \int_a^{\infty} |\psi| dt \ e^{-ax}$$

Letting $a \to \infty$ in the first integral only enhances the inequality, and we have

$$\mathcal{L}(\psi) = o(x^{-(\alpha+1)})$$

Next observing that

$$x = |z| \cos \theta$$

and that $-\pi/2 + \delta \leq \theta \leq \pi/2 - \delta$, gives us the result.

Theorem 223. (Watson's Lemma, 2nd form). For ψ satisfying (2.2.2) and such that in the neighborhood of the origin

(2.2.3) $$\psi(t) = \sum_{n=0}^{N} c_n t^{\gamma_n} + \rho_N(t)$$

with the c_n and the

$$\gamma_n = \alpha_n + i\beta_n$$

constants. Also

$$-1 < \alpha_0 \leq \alpha_1 \leq \alpha_2 \cdots \leq \alpha_N$$

and

$$\rho_N(t) = o(t^{\gamma_N})$$

Then

(2.2.4) $$\mathcal{L}(\psi) = \sum_{n=0}^{N} c_n \gamma_n! \, z^{-(\gamma_n+1)} + r_N(z)$$

with

(2.2.5) $$r_N = o(z^{-\gamma_N-1})$$

for $z \in S^\delta$ and $|z| \to \infty$.

Proof. Since $\mathcal{L}$ is a linear operator, substituting (2.2.3) in $\mathcal{L}$ gives

$$\mathcal{L}(\psi) = \sum_{n=0}^{N} c_n \gamma_n! z^{-(\gamma_n+1)} + \mathcal{L}(\rho_N)$$

The estimate (2.2.5) on

$$r_N = \mathcal{L}(\rho_N)$$

follows from the first form of Watson's Lemma.

The condition (2.2.2) on $\psi(t)$ in Watson's Lemma is clearly too severe, since the AD of $\mathcal{L}(\psi)$ only depends on the behavior of ψ at the origin.

<u>Corollary</u>. If ψ is locally absolutely integrable, satisfies (2.2.3) at the origin and there exists a "b" such that,

$$\psi = O(e^{bt}) \tag{2.2.6}$$

for $t \to \infty$, then (2.2.4), (2.2.5) still applies.

<u>Proof</u>. Consider

$$\mathcal{L}(\psi) = \mathcal{L}(H(t-2b)\psi) + \mathcal{L}([1-H(t-2b)]\psi) \tag{2.2.7}$$

Then

$$\begin{aligned} |\mathcal{L}(H(t-2b)\psi)| &= \left|\int_{2b}^{\infty} e^{-zt}\psi(t)dt\right| \\ &\leq e^{-bx}\int_{2b}^{\infty} e^{-(t-b)x}|\psi(t)|dt \end{aligned}$$

is exponentially small. Watson's lemma applies to the second integral of (2.2.7) and gives us the result.

The restriction to S^{δ} in Watson's lemma can be considerably relaxed if ψ in $\mathcal{L}(\psi)$ is analytic in a sector including the real axis. We do this by appealing to Construction II given in the previous section. Recalling the path P used there, we now define,

$$P_0(\theta) = P(\theta;R=0)$$

i.e. a ray at angle θ from the origin to ∞. Next suppose $\psi(w)$ is analytic in $S(-\alpha,\beta)$ $\alpha,\beta \geq 0$, and $O(e^{\gamma z})$ for $|z| \to \infty$ in that sector. Then for z sufficiently large, the analytic continuation of

$$\Phi(z) = \mathcal{L}(\psi)$$

is given by

$$\Phi(z) = \int_{P_o(\theta)} e^{-zw}\psi(w)dw \tag{2.2.8}$$

where

$$-\frac{\pi}{2} - \theta < \arg z < \frac{\pi}{2} - \theta \tag{2.2.9}$$

and

$$-\alpha < \theta < \beta \tag{2.2.10}$$

The convergence of (2.2.8) for θ in the range (2.2.9) follows from $\psi = O(e^{\gamma w})$ and that (2.2.8) gives the analytic continuation of $\mathcal{L}(\psi)$ follows from the same argument given in Construction II.

Next we assume that as $|w| \to 0$

$$\psi = \sum_{n=o}^{N} c_n w^{\gamma_n} + \rho_N(w), \quad w \in S(-\alpha,\beta) \tag{2.2.11}$$

where c_n and γ_n are as before, and

$$\rho_N = o(w^{\gamma_N}) \tag{2.2.12}$$

On setting

$$w = t\, e^{i\theta}$$

in (2.2.8) we obtain

$$\phi(z) = e^{i\theta}\int_0^\infty e^{-zte^{i\theta}}\psi(te^{i\theta})dt$$

From (2.2.11) we can write

$$\psi(te^{i\theta}) = \sum_{n=0}^{N} c_n(e^{i\theta})^{\gamma_n}t^{\gamma_n} + \hat{\rho}_N(t)$$

From (2.2.12)

$$\hat{\rho}_N(t) = o(t^{\gamma_N})$$

Also it is clear that $\psi(te^{i\theta})$ is exponentially bounded as a function of t. Therefore since

$$|\arg z + \theta| < \frac{\pi}{2}$$

we can apply Watson's lemma and therefore

$$\phi(z) = e^{i\theta}\sum_{n=0}^{N} c_n(e^{i\theta})^{\gamma_n}\gamma_n!\left(\frac{1}{ze^{i\theta}}\right)^{\gamma_n+1} + r_n(z)$$

On clearing terms we obtain (2.2.4), (2.2.5) again. However this was obtained for $\arg z$ such that (2.2.9) and (2.2.10) holds. We have therefore proven:

<u>Theorem 224.</u> (Generalized Watson's Lemma)

In the sector $S(-\alpha,\beta)$, $\alpha,\beta > 0$,

1. $\psi(w)$ analytic
2. $\psi(w) = o(e^{\gamma w})$, $|w| \to \infty$, γ a constant
3. $\psi(w) = \sum_{n=0}^{N} c_n w^{\gamma_n} + \rho_N(w)$, as $|w| \to 0$, $w \in S(-\alpha,\beta)$,

   $\rho_N = o(w^{\gamma_N})$
4. $\gamma_n = \alpha_n + i\beta_n$, $-1 < \alpha_0 \leq \alpha_1 \leq \dots \leq \alpha_N$.

Then

(2.2.13) $$\mathcal{L}(\psi) = \sum_{n=o}^{N} \gamma_n! c_n z^{-\gamma_n - 1} + r_N(z)$$

for $|z| \to \infty$

(2.2.14) $$z \in S(-\frac{\pi}{2} - \beta + \delta, \frac{\pi}{2} + \alpha - \delta), \quad \delta > 0$$

with

(2.2.15) $$r_N(z) = o(z^{-\gamma_N - 1})$$

The size of the sector (2.2.14) follows from (2.2.9, 10).

Exercise 22. Find the AD of

$$\int_o^1 e^{-\frac{z}{t} + t + zt} dt$$

for $|z| \to \infty$. In what sector is the AD valid?

Exercise 23. Consider

$$\int_o^1 \frac{e^{-zt^n}}{1+t} dt$$

n, integer > 0, for z large. Specify the sector of validity of the AD.

Exercise 24. Obtain the AD of

$$\int_o^1 e^{-sz} \tan h\ s\ ds$$

for $|z| \to \infty$. Specify sector of validity.

Exercise 25. Find the asymptotic expansion of

$$\int_0^1 t^z \cos^2(1-t)dt$$

for $\operatorname{Re} z > 0$ and $|z| \to \infty$.

We point out, that by virtue of Construction II, it is only necessary to consider $\mathcal{L}(\psi)$ for $|\arg z| < \pi$ (or its equivalent). The analytic continuation for other values of $\arg z$ follows from (2.1.4) and accordingly so does the appropriate AD.

Example. Consider the Laplace transform of $(1+w^2)^{-1}$, or in complex notation

$$\phi(z) = \int_0^\infty \frac{e^{-zw}}{1+w^2} dw$$

The function $(1+w^2)^{-1}$ is analytic everywhere except at $w = \pm i$ where it has simple poles. Therefore

$$\psi = (1+w^2)^{-1} \sim \sum_{n=0}^{\infty} (-)^n w^{+2n}$$

as $|w| \to 0$ $w \in S^\delta$. From the generalized Watson's lemma, therefore

$$\phi(z) \sim \sum_{n=0}^{\infty} (-)^n (2n)! \; z^{-2n-1}$$

$$|z| \to \infty \quad |\arg z| < \pi$$

From (2.1.4) we can write

$$\phi(z) - \phi(ze^{-i2\pi}) = [\phi] = -\oint \frac{e^{-zw}}{1+w^2} dw$$

$$= -2\pi i\left\{\frac{e^{-iz}}{2i} - \frac{e^{iz}}{2i}\right\} = \pi e^{iz} - \pi e^{-iz}.$$

From this we obtain the AD of Φ for all arg z, and we notice that both the negative and positive imaginary axes are Stokes lines. In particular

$$\Phi(z) \sim + \pi e^{iz} \qquad \pi < \arg z < 2\pi$$

$$\Phi(z) \sim - \pi e^{-iz} \qquad 2\pi < \arg z < 3\pi$$

and so forth.

This example illustrates another piece of folklore which is useful to know. The rays $-(\beta + \frac{\pi}{2})$ and $(\alpha + \frac{\pi}{2})$, (2.1.14) which appear in the conclusion of the generalized Watson's lemma are usually Stokes lines for the $\mathcal{L}(\psi)$ under study. The reason for this becomes clear when we recognize that the $-\alpha$ and β appear as limits for the asymptotic expansion of ψ as $|w| \to 0$, because of singularities, such as poles or branch cuts, of ψ. Then in analytically continuing $\mathcal{L}(\psi)$ the contribution of these singularities is picked-up and felt as exponential growth an angle $\pi/2$ later.

Exercise 26. The Hankel function $H_0^{(1)}(z)$ is given by

$$H_0^{(1)}(z) = \frac{2e^{-iz}i}{\pi} \int_0^{\infty} \frac{e^{-zs}ds}{(s^2+2is)^{\frac{1}{2}}}$$

(the branch of the radical is such that it approaches s for $s \to \infty$). Find $H_0^{(1)}(z)$ for $|z| \to \infty$ for all arguments of z. Indicate Stokes lines.

Exercise 27. Noting that

$$\operatorname{erfc} z = \frac{2}{\sqrt{\pi}} \int_z^{\infty} \exp(-t^2)dt$$

$$= \frac{2}{\sqrt{\pi}} e^{-z^2} \int_0^{\infty} e^{-2zs}e^{-s^2} ds$$

Find the AD of erfc z for all values of arg z. Indicate Stokes lines.

Exercise 28. Consider

$$\mathscr{G}(z) = \int_0^\infty \frac{e^{-t}\,dt}{1 + \frac{t}{z}}$$

and obtain the AD for all arg z. Compare with earlier development given on page 59.

Fourier Integrals

If $z = iy$, y real, in (2.2.1) we obtain an example of a Fourier integral. Because of their natural importance we wish to obtain results describing Fourier integrals at infinity. Several such results have already been obtained. In section 2.1 (2.1.11) we obtained the asymptotic expansion when ψ, the untransformed function is differentiable. Results of wider application follow if ψ is analytic in a sector including the positive real axis. This follows from the Generalized Watson's lemma given above. When the function $\psi(t)$ is not analytic, the situation is delicate. Unlike the Laplace transform the Fourier transform does not in general depend on only the behavior of ψ at the origin. This is signaled by (2.1.11) which shows that both endpoints of integration can be of equal importance. Of even more importance is the smoothness of $\psi(t)$ in the interior. Exercise 12c, demonstrated that a singularity in the interval of integration can contribute at a lower order than the simple endpoint contribution.

We first make some preliminary remarks. We consider integrals of the form

$$\int_\alpha^\beta e^{ity}\,\hat{\psi}(t)dt \tag{2.2.16}$$

for t and y real and $|y| \to \infty$. It is of course assumed that ψ is integrable. Further it will be assumed that $\hat{\psi}$ is sufficiently smooth (the meaning of this, in the present context, will emerge below) except at a finite number of interior points, say $\alpha_1, \alpha_2, \ldots, \alpha_n$. We can therefore write

$$\int_\alpha^\beta e^{ity}\,\hat{\psi}(t)dt = \sum_{i=0}^{n} \int_{\alpha_i}^{\alpha_{i+1}} e^{ity}\,\hat{\psi}(t)dt$$

with $\alpha_0 = \alpha$, $\alpha_{n+1} = \beta$, and consider each such integral individually. Hence we may assume $\hat{\psi}(t)$ in (2.2.16) to be smooth in the interior of the interval of integration. By doing this we have arranged to have the AD to depend only on the behavior of $\hat{\psi}(t)$ at the endpoints. We will avoid the nuisance of dealing with both endpoints by introducing the function

$$\nu(t) = \frac{\int_t^\beta e^{-\frac{1}{u-\alpha}+\frac{1}{u-\beta}}du}{\int_\alpha^\beta e^{-\frac{1}{u-\alpha}+\frac{1}{u-\beta}}du} \tag{2.2.17}$$

Such functions were introduced by van der Corput, who called them neutralizers. We see that

$$\nu(\alpha) = 1$$

$$\nu(\beta) = 0$$

and

$$\left.\frac{d^n}{dt^n}\nu\right|_{\alpha,\beta} = -\frac{d^{n-1}}{dt^{n-1}}\left.\frac{e^{-\frac{1}{t-\alpha}+\frac{1}{t-\beta}}}{\int_\alpha^\beta e^{-\frac{1}{u-\alpha}+\frac{1}{u-\beta}}du}\right|_{\alpha,\beta} = 0 \tag{2.2.18}$$

for all $n \geq 1$ as α, β are approached from the interior. If ν is referred to as a neutralizer for the point α, then $1-\nu$ is a neutralizer for the point β. In a manner of speaking $\nu(t)$ is identically unity at $t = \alpha$, at least in the sense that for all meaningful n,

$$\left.\frac{d^n}{dt^n}(\nu\hat{\psi})\right|_{t=\alpha} = \left.\frac{d^n}{dt^n}\hat{\psi}\right|_{t=\alpha} \tag{2.2.19}$$

This holds for those n such that the right hand side has a meaning. Actually the form of the neutralizer is of no consequence. We only need the existence of functions having its properties.

Next we write

$$\int_\alpha^\beta e^{ity}\,\hat{\psi}dt = \int_\alpha^\beta e^{ity}\,\hat{\psi}\nu(t)dt + \int_\alpha^\beta e^{ity}\,\hat{\psi}(1-\nu)dt$$

This accomplishes our goal of isolating each of the endpoints. Defining

$$\nu(t) \equiv 0, \quad t \geq \beta$$
$$\hat{\psi}(t) \equiv 0, \quad t > \beta$$

we may consider simply

$$\int_\alpha^\infty e^{ixt}\,\tilde{\psi}(t)dt$$

where it is understood that $\tilde{\psi} \equiv 0$ for $t \geq \beta$ and that $\tilde{\psi} = \hat{\psi}\nu$ is as smooth as the original $\hat{\psi}$. Finally setting $t = s+\alpha$

$$\int_\alpha^\infty e^{itx}\,\tilde{\psi}(t)dt = e^{i\alpha x}\int_0^\infty e^{isx}\,\tilde{\psi}(s+\alpha)ds$$
$$= e^{i\alpha x}\mathscr{F}(\psi).$$

Therefore we may consider without loss of generality

$$\mathscr{F}(\psi) = \int_0^\infty e^{ixt}\psi(t)dt \tag{2.2.20}$$

under the assumption that with the exception of the origin $\psi \in C_0^\infty$. Actually the last will be considerably relaxed in what follows.

The most general results on the asymptotic behavior of Fourier transforms were obtained by Erdelyi [see J. Sci. Indust. Appl. Math. 3, 17, 1955]. We are indebted to him for demonstrating how far we may go with so elementary a method as parts integration. In the following we shall use a different approach which produces a somewhat sharper result.

Lemma. $\phi \in C_0^N$, $\phi^{(n)}(t=0) = 0$, $n=0,\ldots,N-1$. Further let γ be such that $0 \geq$

Re $\gamma > -1$, then $t^{\gamma}\phi$ has N absolutely integrable derivatives and

$$\frac{d^{(n)}}{dt^n}(t^{\gamma}\phi)\Big|_{t=0} = 0 \qquad n = 0,\ldots,N-1.$$

Proof. To prove this we use an induction argument. First for $N = 0$,

$$t^{\gamma}\phi\Big|_{t=0} = 0$$

and for $N = 1$

$$\frac{d}{dt}(t^{\gamma}\phi) = t^{\gamma}\phi' + \gamma t^{\gamma}\frac{\phi}{t}$$

which is clearly integrable. Next consider $N = N$

$$\frac{d}{dt}(t^{\gamma}\phi) = t^{\gamma}[\phi' + \gamma\frac{\phi}{t}].$$

Therefore by induction since $\phi' + \gamma\phi t^{-1}$ has $N-1$ derivatives the first $N - 2$ of which are zero at $t = 0$ the lemma is proven.

Theorem 2.25. Under the hypothesis of the above lemma for $\psi = t^{\gamma}\phi$

$$\mathcal{L}(\psi) = o(z^{-N})$$

$|z| \to \infty$ and Re $z \geq 0$.

Proof. Parts integrating N-times

$$\mathcal{L}(\psi) = \frac{1}{z^N}\int_0^{\infty} e^{-zt}\frac{d^N}{dt^N}(t^{\gamma}\phi)dt$$

Since $\frac{d^N}{dt^N}(t^{\gamma}\phi)$ is absolutely integrable by the lemma, the Riemann-Lebesgue lemma proves our statement.

<u>Corollary</u>. If $\phi \in C_0^\infty$ and $\phi \equiv \phi(0)$ in the neighborhood of the origin

(2.2.21) $$\mathcal{L}(t^{\gamma+N}\phi) = \frac{(\gamma+N)!\,\phi(0)}{z^{\gamma+N+1}} + o(z^{-\infty})$$

where $0 \geq \mathrm{Re}\,\gamma > -1$ and $o(z^{-\infty})$ signifies a quantity vanishing to all orders.

<u>Proof</u>. Parts integrating N-times and using the above lemma we have

$$\mathcal{L}(t^{\gamma+N}\phi) = \frac{(\gamma+N)!}{\gamma!\,z^N}\mathcal{L}(t^\gamma\phi) + o(z^{-\infty})$$

If $0 > \mathrm{Re}\,\gamma > -1$ we rewrite this as

$$\mathcal{L}(t^{\gamma+N}\phi) = \frac{(\gamma+N)!\,\phi(0)}{z^{\gamma+N+1}} + \frac{(\gamma+N)!}{\gamma!\,z^N}\mathcal{L}(t^\gamma[\phi-\phi(0)]) + o(z^{-\infty})$$

and $(\phi-\phi(0)) \in C_0^\infty$ are identically zero at $t = 0$, Theorem <u>225</u> gives the result. If $\mathrm{Re}\,\gamma = 0$, we must first restrict attention to $\mathrm{Re}\,z > 0$ and then use a continuity argument.

<u>Theorem 226</u>. For $\psi = t^\gamma\phi$, $0 \geq \mathrm{Re}\,\gamma > -1$, $\phi \in C_0^N$

(2.2.22) $$\mathcal{L}(\psi) = \sum_{n=0}^{N-1} \frac{\phi^{(n)}(0)(\gamma+n)!}{n!\,z^{n+\gamma+1}} + o(z^{-N})$$

for $|z| \to \infty$ $\mathrm{Re}\,z \geq 0$.

<u>Proof</u>. Let $\nu(t)$ represent a function which is identically unity on the support of ϕ and zero elsewhere. Therefore

$$\mathcal{L}(\psi) = \mathcal{L}(\nu\psi)$$

$$\mathcal{L}(\nu\psi) = \int_0^\infty t^\gamma \phi \nu \, e^{-zt} dt$$

$$= \int_0^\infty t^\gamma \nu \sum_{n=0}^{N-1} \frac{\phi^{(n)}(0)t^n}{n!} e^{-zt} dt$$

$$+ \int_0^\infty t^\gamma \nu \{\phi(t) - \sum_{n=0}^{N-1} \frac{\phi^{(n)}(0)t^n}{n!}\} e^{-zt} dt$$

$$= \sum_{n=0}^{N-1} \frac{\phi^{(n)}(0)(\gamma+n)!}{n! z^{\gamma+n+1}} + o(z^{-\infty})$$

$$+ \int_0^\infty t^\gamma \{\phi(t) - \sum_{n=0}^{N-1} \frac{\phi^{(n)}(0)t^n}{n!}\} \nu e^{-zt} dt$$

The term in the braces is C_0^N and has N-1 zero derivatives at the origin therefore applying Theorem 215 gives us the required result. Piecing together the various transformations and reductions with Theorem 226, we may

Exercise 29. Prove the following theorem due to Erdelyi.

Theorem 227. If $-1 < \text{Re } \lambda, \text{Re } \mu \le 0$, and $\phi(t)$ is N times continuously differentiable for $\alpha \le t \le \beta$, then

$$\int_\alpha^\beta e^{ixt}(t-\alpha)^\lambda(\beta-t)^\mu \phi(t)dt = B_N(x) - A_N(x) + o(x^{-N})$$

where

$$A_N(x) = \sum_{n=0}^{N-1} \frac{(\lambda+n)! \, e^{\frac{i\pi}{2}(n+\lambda-1)}}{n! \; x^{n+\lambda+1}} \cdot e^{ix\alpha}[\frac{d^n}{dt^n}\{(\beta-t)^\mu \phi(t)\}]_{t=\alpha}$$

$$B_N(x) = \sum_{n=0}^{N-1} \frac{(\mu+n)! \, e^{\frac{i\pi}{2}(n+\mu-1)}}{n! \; x^{n+\mu+1}} \quad e^{ix\beta}[\frac{d^n}{dt^n}\{(t-\alpha)^\lambda \phi(t)\}]_{t=\beta}.$$

The inclusion of a logarithmic singularity may also be included in the analysis. For this analysis consult, A. Erdelyi, J. Soc. Indust. Appl. Math. Vol. 2, No. 1, 38, 1956.

2.3. Laplace's Formula and Its Generalization.

In this section we consider

$$(2.3.1) \qquad L(t) = \int_\alpha^\beta e^{+th(s)} g(s)ds$$

for $t \to \infty$ and h real. The classical result for such integrals is due to Laplace who argues that its main contribution comes from the neighborhood of the global maximum of $h(s)$, say s_o. Assuming that $h(s)$ is twice differentiable, this leads to

$$(2.3.2) \qquad \begin{aligned} L(t) &\sim \int_{-\infty}^{\infty} e^{+t(h(s_o) + \frac{h''(s_o)(s-s_o)^2}{2})} g(s_o)ds \\ &= g(s_o)e^{+th(s_o)}\left(\frac{2\pi}{-th''(s_o)}\right)^{\frac{1}{2}} \end{aligned}$$

(since s_o is the maximum, $-h''(s_o) > 0$). This is referred to as Laplace's formula.

The method for verifying (2.3.2) as well as obtaining subsequent terms in the AD, is simply to reduce (2.3.1) to a Laplace transform and then employ Watson's lemma. We will also at the same time treat situations which fall outside the realm of Laplace's formula, e.g. if either or both $h''(s_o)$ and $g(s_o)$ do not exist. We present the results in the form of a construction instead of a theorem and formula since the former is more useful in practice.

Asymptotic Reduction of $\int_\alpha^\beta g(s)\exp[t\, h(s)]ds$ to Laplace Transforms

Without loss of generality we assume that $h(s)$ has a single global maximum for $s_o \in [\alpha,\beta]$, which may or may not be an endpoint.

Also we assume there exists a neighborhood of s_o so that $h(s)$ decreases monotonically from $h(s_o)$.

Next since $h_o = h(s_o)$ is a global maximum there exist $\varepsilon > 0$ and $\delta > 0$ such that

$$h_o - h(s) > \varepsilon \quad \text{for} \quad |s-s_o| > \delta$$

Moreover let these be chosen so that $h(s)$ is monotonically decreasing to the right and left of s_o. We then write

$$L(t) = e^{th_o} \int_{|s-s_o|<\delta} e^{t(h-h_o)} g(s)ds$$
$$+ e^{th_o} \int_{\substack{|s-s_o|\geq\delta \\ s\in(\alpha,\beta)}} e^{t(h-h_o)} g(s)ds$$

Assuming the absolute integrability of g, the second integral of the right hand is $O(e^{t(h_o-\varepsilon)})$ and we have

$$L = e^{th_o} \int_{s_o}^{s_o+\delta} e^{t(h-h_o)} g(s)ds + e^{th_o} \int_{s_o-\delta}^{s_o} e^{t(h-h_o)} g(s)ds$$
$$+ O(e^{t(h_o-\varepsilon)}).$$

From the monotonicity assumption we know that the transformation

(2.3.3) $$\sigma = h_o - h(s)$$

is invertible. Denote the inverse by

$$s = s^{\pm}(\sigma)$$

where $s^+(\sigma)\uparrow$ for $\sigma\uparrow$ is to be used in the first integral and $s^-(\sigma)\downarrow$ for $\sigma\uparrow$ is to be used in the second integral. Then writing

$$\tau^{\pm} = h_o - h(s_o \pm \delta) > 0$$

(if s_o is an endpoint one of these should be zero) we have

$$L = e^{th_o} \int_0^{\tau+} e^{-\sigma t} g(s^+(\sigma)) \frac{ds^+}{d\sigma} d\sigma$$

$$(2.3.4) \qquad - e^{th_o} \int_0^{\tau-} e^{-\sigma t} g(s^-(\sigma)) \frac{ds^-}{d\sigma} d\sigma$$

$$+ O(e^{t(h_o-\varepsilon)})$$

where it is assumed that $\frac{ds^\pm}{d\sigma}$ have a meaning.

At this point the expansion of

$$(2.3.5) \qquad g(s^\pm(\sigma)) \frac{ds^\pm}{d\sigma}$$

in the neighborhood of the origin is required. From this the asymptotic expansion is obtained from Watson's lemma. One additional observation is that the restriction to real t may be now dropped and we instead require $t \in S^\delta$.

An important point to note is that no smoothness requirements on either $h(s)$ or $g(s)$ in obtaining (2.3.4).

As a specific case let us assume that $h(s) \in C^2$ for $s = s_o$, and that

$$g = g_o(s-s_o)^\alpha + o((s-s_o)^\alpha)$$

Re $\alpha > -1$. Then the transformation (2.3.3) in the neighborhood s_o is

$$\sigma = h_o - [h_o + \frac{(s-s_o)^2}{2} h_o'' + o((s-s_o)^2)].$$

Then from Theorem 123 on the inversion of asymptotic expansions,

$$s^\pm = s_o \pm (\frac{2\sigma}{-h_o''})^{\frac{1}{2}} + o(\sigma^{1/2})$$

and

$$\frac{ds^{\pm}}{d\sigma} \sim \pm\left(\frac{1}{-2\sigma h''_o}\right)^{\frac{1}{2}}$$

Note that the branches $s^{\pm}$ were chosen to correspond to the two integrals of L, (2.3.4) according to the above discussion. In our case (2.3.5) therefore becomes

$$g(s^{\pm})\ \frac{ds^{\pm}}{d\sigma} = \pm\ g_o[\pm 2/h''_o]^{\alpha/2}(-2h'')^{-1/2}\sigma^{(\alpha-1)/2} + o(\sigma^{(\alpha-1)/2})$$

Hence from Watson's lemma

$$\text{(2.3.6)} \qquad L \sim (-2h''_o)^{-1/2} e^{th_o} g_o\{(-2/h''_o)^{\frac{\alpha}{2}} + (2/h''_o)^{\frac{\alpha}{2}}\}(\tfrac{\alpha}{2} - \tfrac{1}{2})!\, t^{-(\alpha+1)/2}$$

In particular if $\alpha = 0$, we obtain Laplace's formula

$$L \sim g_o\ e^{th_o}\left(\frac{2\pi}{-h''_o t}\right)^{\frac{1}{2}}$$

(We have used $(-\frac{1}{2})! = \sqrt{\pi}$.)

As a counterpart to the Laplace form in the general case we suppose that the global maximum of $h(s)$ in (2.3.1) is at $s = \alpha$, and that

$$\text{(2.3.7)} \qquad \begin{aligned} h(s) &= h(\alpha) - \eta(s-\alpha)^{\mu} + o((s-\alpha)^{\mu}) \\ g(s) &= g(\alpha)(s-\alpha)^{\nu} + o((s-\alpha)^{\nu}) \end{aligned}$$

with $\eta,\mu > 0$ and $\nu > -1$. Then by the above discussion, and setting $w = s-\alpha$,

$$L \sim g(\alpha)e^{th(\alpha)} \int_0^{\infty} e^{-\eta t w^{\mu}} w^{\nu} dw$$

or on changing variables

(2.3.8) $$L \sim g(\alpha)e^{th(\alpha)}\left(\frac{\nu+1-\mu}{\mu}\right)!\ \frac{(\eta t)^{-(\nu+1)/\mu}}{\mu}$$

Exercise 30: Demonstrate (2.3.8) under (2.3.7)

Example. Consider as $n \to \infty$, the following integral ,

$$I_n = \int_a^b [p(x)]^n dx$$

with $p \in C^2$ and such that $p(x_o) > p(x)$, $x_o(\neq x) \in [a,b]$. By Laplace's formula we immediately have

$$I_n = \int_a^b e^{n\,\ell n\,p(x)} dx \sim e^{n\,\ell n\,p(x_o)}\left(\frac{-2\pi\,p_o}{n\,p_o''}\right)^{\frac{1}{2}}$$

$$\left[\frac{d}{dx}\,\ell n\,p\Big|_{x_o} = \frac{p'}{p}\Big|_{x_o} = 0,\quad p(x_o) \neq 0\right]$$

Noting that $[\frac{a}{n}]^{1/n} \to 1$, we also see

$$\lim_{n \to \infty} [I_n]^{1/n} = p(x_o)$$

or L_∞ is the maximum norm.

Example. Consider $s!$ for $s \to \infty$

$$s! = \int_0^\infty e^{-t} t^s dt = \int_0^\infty e^{-t+s\ell nt} dt$$

Set $t = sp$

$$s! = s\int_0^\infty e^{-sp+s\ \ell ns+s\ell np} dp$$

$$= s^{s+1}\int_0^\infty e^{-s(p-\ell np)} dp \sim s^{s+1} e^{-s}\left(\frac{2\pi}{s}\right)^{\frac{1}{2}} ,$$

Stirling's formula.

Exercise 30. Find the leading term for $n \to \infty$ in the AD of

(a) $\int_0^{\pi} x^n \sin x \, dx$ (b) $\int_0^1 \frac{e^x x^n}{(1+x^2)^n} dx$

(c) $\int_0^1 t^n \sin^2 t \, dt$ (d) $\int_0^{\infty} \frac{e^{nt}}{t^t} dt$

Exercise 31. Find the leading term in the AD as $s \to \infty$ of

(a) $\int_0^1 e^{s \frac{\sin t}{t}} \frac{dt}{\sqrt{t}}$

(b) $\int_0^{\frac{\pi}{2}} e^{-st^2 \sin t} t dt.$

In actual practive it is tedious to carry through all the transformations which ultimately reduce the integral (2.3.1) to a Laplace transform. For this reason we mention the following formal procedure for the asymptotic evaluation of integrals of type (2.3.1). The above rigorous discussion justifies the following recipe.

Again we represent the global maximum by s_0, and in order to avoid minor technicalities we suppose that it is located at the lower limit of integration

(2.3.9) $$\int_{s_0}^{\beta} e^{th(s)} g(s) ds$$

We represent the expansions around s_0 as follows,

$$h = h_0 - a(s-s_0)^{\beta} + r(s-s_0)$$
$$g = \sum_{i=0}^{N} (s-s_0)^{\gamma_i} \beta_i + o((s-s_0)^{\gamma_n})$$

with

$$a > 0, \ \beta > 0$$

$$-1 < \mathrm{Re}\ \gamma_0 \leq \mathrm{Re}\ \gamma_1 \leq \dots \leq R\gamma_N$$

and

$$r(x) = o(x^{\beta})$$

Then after setting $(s-s_0) = x$

(2.3.10)
$$\int_{s_0}^{\beta} e^{th(s)} g(s)ds \sim \int_o^{\infty} e^{-ax^{\beta}t}[1-atr(x) + \frac{at^2}{2} r^2(x) + \dots] \sum_{i=0} \beta_i x^{\gamma_i} dx$$

A subsequent expansion of $r(x)$, $r^2(x)$, $r^3(x)$, etc., is necessary for the evaluation. The following formula

(2.3.11)
$$\int_o^{\infty} e^{-ax^{\beta}t} x^{\gamma} dx = \frac{(\frac{\gamma+1-\beta}{\beta})!}{\beta(at)^{\frac{\gamma+1}{\beta}}}$$

then allows an explicit evaluation.

Exercise 32. By retracing the steps leading to (2.3.4) verify that (2.3.10) is rigorously the AD of (2.3.9).

2.4. Kelvin's Formula and Generalizations

In this section we consider integrals of the type

(2.4.1)
$$\int_{\alpha}^{\beta} e^{ith(s)} g(s)ds$$

for t real and $t \to \infty$. For monotonic differentiable $h(s)$ this case has been treated by (2.1.13). Let us assume that h has a single stationary point $s_o \in (\alpha,\beta)$,

$$h'(s_o) = 0, \quad h''(s_o) \neq 0$$

In certain superficial aspects the argument for obtaining the leading term of (2.4.1) resembles that used in obtaining Laplace's Formula (2.3.2).

Our understanding of integrals such as (2.4.1) as well as the simpler transforms (2.2.16) tells us that these vanish as $t \to \infty$. The underlying reason is that on the scale of oscillation of $e^{ith(s)}$ as $t \to \infty$ the function $g(s)$ appears as a constant and hence the contributions of successive oscillations effectively cancel. If however $h(s)$ is locally zero (or equivalently a constant) we might expect this effect to be ameliorated. It is therefore plausible that

$$\int_\alpha^\beta e^{ith(s)} g(s)ds \sim e^{ih(s_0)t} \int_{-\infty}^{\infty} g(s_0)\exp[\frac{ih''(s_0)(s-s_0)^2 t}{2}]ds$$

On carrying out the integration we obtain Kelvin's formula,

$$\int_\alpha^\beta e^{ih(s)t} g(s)ds \sim g(s_0)\left(\frac{2\pi}{t|h''(s_0)|}\right)^{\frac{1}{2}} e^{i[h(s_0)t + \frac{\pi}{4}\frac{h''(s_0)}{|h''(s_0)|}]}$$

The analogy with the derivation of Laplace's formula is entirely superficial. For it is clear that the discussion does not depend on whether the stationary point is a maximum or minimum. Also we recall that an endpoint contributes $O(\frac{1}{t})$. This lends support to the above derivation since Kelvin's formula contributes at the $O(t^{-\frac{1}{2}})$, but it also states that at $O(t^{-1})$ (the next order), the endpoint contribution must be retained. Recall that in the case of the extension of Laplace's formula all orders come from the neighborhood of the maximum point.

Our treatment of integrals of the form (2.4.1) is simply to reduce them to Fourier integrals. In this sense we parallel the discussion reducing integrals of the form (2.3.1) to Laplace integrals. The one important difference lies in recognizing that discontinuities of $h(s)$ and $g(s)$ and their derivatives, end-points and stationary points of $h(s)$ will now play a role in the full asymptotic development of (2.4.1). Any point $s_0 \in [\alpha,\beta]$ where one of the above possibilities occur will be referred to as a critical point of (2.4.1).

Let the critical points of (2.4.1) be represented by

$$\alpha = \alpha_0 < \alpha_1 < \alpha_2 < \alpha_3 \dots < \alpha_n = \beta$$

I.e., at an interior point of (α_k, α_{k+1}) the following is true,

1. $h(s)$ is sufficiently smooth.
2. $g(s)$ is sufficiently smooth.
3. $h'(s) \neq 0$.

We may therefore consider (2.4.1) under the assumption that $h(s)$ is monotonic in the interval, and that $h(s)$ and $g(s)$ are sufficiently smooth in the interval. As we have seen the meaning of smoothness varies with our needs. In fact the above subdivision will change with the degree of smoothness required.

In the usual way we transform to the origin and introduce a neutralizer so that we consider

(2.4.2) $$\mathscr{I}(t) = \int_0^\beta e^{ith(s)}\, \hat{g}(s)\,ds$$

where β is not usually the same as the above. By virtue of the supposed neutralizer $\hat{g}(\beta) \equiv 0$, in the sense that it and all its pertinent derivatives are zero at the point β. From the above discussion we regard $h(s)$ as being monotonic in $(0,\beta)$. Therefore the sole critical point is the origin.

We next write,

$$h(s) = h_0 + s^\rho u(s)$$

$$\hat{g}(s) = s^\gamma v(s)$$

with

$$u(0), \quad v(0) \neq 0$$

$$\rho > 0$$

$$0 \geq \mathrm{Re}\ \gamma > -1.$$

Also we take

$$u'(s), v(s) \in C^N$$

for if $u'(s) \in C^p$ and $v(s) \in C^q$ then $N = \min(p,q)$. The restriction to real t may be lifted and we substitute $-z$ for it with the condition

$$\mathrm{Re}\ z \geq 0.$$

Also no loss of generality results from taking $h' > 0$. With these remarks our integral takes the form

(2.4.3) $$\mathcal{J}(z) = e^{-zh_0} \int_0^\beta e^{-zs^\rho u(s)}\ s^\gamma v(s) ds$$

We next transform by writing

(2.4.4) $$s^\rho u(s) = x^\rho$$

with the branch chosen so that $x\uparrow$ for $s\uparrow$. Under this transformation,

(2.4.5) $$\mathcal{J}(z) = e^{-zh_0} \int_0^P e^{-zx^\rho} x^\gamma\ k(x) dx$$

with P such that $\beta^\rho u(\beta) = P^\rho$, and

(2.4.6) $$k(x) = [\frac{s(x)}{x}]^\gamma v(s(x))\frac{ds}{dx}$$

where $s(x)$ denotes the appropriate branch of (2.4.4). Next we show $k \in C^N$ and $k(P) \equiv 0$. Differentiating (2.4.4) we easily find,

$$\frac{ds(x)}{dx} = \frac{x^{\rho-1}}{s^{\rho-1}}[\frac{\rho}{\rho u(s) + s\frac{du}{ds}}] = \frac{[u(s)]^{1-\frac{1}{\rho}}\rho}{\rho u(s) + s\frac{du}{ds}}.$$

Hence since $u(0) \neq 0$, $\frac{ds}{dx}$ has N derivatives. Moreover since

$$k(x) = \frac{\rho v(s(x))[u(s)]^{-\frac{1}{\rho}+1-\frac{\gamma}{\rho}}}{\rho u(s) + s\frac{du}{ds}}$$

it has N derivatives. Finally since $v(s(P)) = v(\beta) \equiv 0$ it follows that $k(P) \equiv 0$.

If $\rho = 1$ no further work is necessary since the evaluation is then reduced to the results of section (2.3). When $\rho \neq 1$ a further reduction is required. To accomplish this we first prove the following analogue to Theorem 215.

Lemma. Consider

$$J(z) = \int_0^\infty e^{-zx^\rho} g(x)dx, \quad \rho > 0$$

g of compact support and having N absolutely integrable derivatives and $g = O(x^{N+\gamma})$ (γ as above) for $x \to 0$, so that $\left.\frac{d^i g}{dx^i}\right|_{x=0} = 0$, $i = 0, N-1$. Then for $\operatorname{Re} z \geq 0$ and $|z| \to \infty$,

$$J(z) = \begin{cases} o(z^{-[\frac{N+\operatorname{Re}(\gamma+1)}{\rho}]}), & \rho \geq 1 \\ o(z^{-N}), & \rho \leq 1 \end{cases}$$

where $[\alpha]$ represents the integer part of α.

Proof. By successive parts integration we find

$$\int_0^\infty e^{-zx^\rho} g(x)dx = \sum_{j=0}^{n-1} \frac{1}{(\rho z)^{j+1}}[(x^{1-\rho}\frac{d}{dx})^j(x^{1-\rho}\ g)]_{x=0}$$

$$+ \frac{1}{(\rho z)^n}\int_0^\infty e^{-zx^\rho}(\frac{d}{dx}\ x^{1-\rho})^n\ g\ dx$$

Clearly the first term vanishes and the second exists if $n \leq N$ is such that

$$N + \text{Re}\ \gamma - n\rho > -1$$

Therefore if $\rho \geq 1$ we take

$$n = [\frac{N+\text{Re}\ \gamma+1}{\rho}]$$

and if $\rho \leq 1$ we take

$$n = N$$

The lemma then follows by recognizing that

$$\lim_{\substack{|z|\to\infty \\ \text{Re}\ z\geq 0}} \int_0^\infty e^{-zx^\rho}(\frac{d}{dx}\ x^{1-\rho})^n g(x)\ dx = 0$$

which follows by setting $x^\rho = y$ and employing the Riemann-Lebesgue lemma.

A sharper estimate verifies that the integer value operator may be dropped - and we use the lemma in that form.

Now we return to the integral (2.4.5). Letting $\nu \in C_0^\infty$ and $\nu(x) = 1$ for $x \in [0,\rho]$, we write $\mathscr{I}$ as

$$\mathscr{I} = \int_0^\infty e^{-zx^\rho} x^\gamma \sum_{i=0}^{N} \frac{k_0^{(i)}x^i}{i!}\ \nu(x)dx\ e^{-h_0 z}$$
$$+ \int_0^\infty e^{-z\gamma^\rho} x^\gamma\{k(x) - \sum_{i=0}^{N} \frac{k_0^{(i)}x^i}{i!}\}\nu(x)dx\ e^{-h_0 z}$$

with $k_o^{(i)} = \frac{d^i}{dx^i} k(x)\Big|_{x=o}$. From the lemma of Theorem 215 and the above lemma the second term is

$$\begin{cases} O(z^{-[\frac{N+\text{Re}\ \gamma+1}{\rho}]}) & \rho \geq 1 \\ O(z^{-N}) & \rho \leq 1 \end{cases}$$

Setting $x^\rho = y$ in the first term

$$\int_o^\infty e^{-zx^\rho} \sum_{i=o}^{N} \frac{k_o^{(i)}}{i!} x^{i+\gamma} \nu(x) dx\, e^{-h_o z}$$

$$= \int_o^\infty e^{-zy} \sum_{i=o}^{N} \frac{k_o^{(i)}}{i!} (y^{\frac{\gamma+i}{\rho}} \hat{\nu}(y) \frac{dx}{dy} dy) e^{-h_o z}$$

$$= \int_o^\infty e^{-zy} \sum_{i=o}^{N} \frac{k_o^{(i)}}{\rho i!} y^{\frac{i+\gamma+1-\rho}{\rho}} \hat{\nu}(y) dy\; e^{-h_o z}$$

where $\hat{\nu}$ has the same properties as ν.

Using (2.2.4) we find that this becomes,

$$e^{-h_o z} \sum_{i=o}^{N} \frac{(\frac{i+\gamma+1-\rho}{\rho})!\; k_o^{(i)}}{\rho i!\; z^{(i+\gamma+1)/\rho}} + o(z^{-\infty})$$

for $\text{Re}\ z \geq 0$, $|z| \to \infty$. This then demonstrates

Theorem 240. For

(2.4.7) $$\phi(z) = \int_o^\beta e^{-zh(s)} \hat{g}(s) ds$$

with

$$h(s) = h_o + s^\rho u(s), \quad u(s) \text{ monotonic}, \ \rho > 0$$

(2.4.8) $$g(s) = s^\gamma v(s), \quad 0 \geq \text{Re}\ \gamma > -1$$

$$u'(s), v(s) \in C^N, \ u(0) \neq 0, \ v(0) \neq 0, \ g(\beta) \equiv 0$$

Then

$$\mathscr{I}(z) = \frac{e^{-zh_o}}{\rho} \sum_{i=o}^{N} \frac{\left(\frac{i+\gamma+1-\rho}{\rho}\right)!\,k_o^{(i)}}{i!\; z^{(i+\gamma+1)/\rho}} + \begin{cases} o\left(z^{-\frac{N+\gamma+1}{\rho}}\right) & \rho > 1 \\ o(z^{-N}) & \rho \le 1 \end{cases} \tag{2.4.9}$$

$$\text{for } \operatorname{Re} z \ge 0, \quad |z| \to \infty$$

where

$$k(x) = \frac{\rho[u(s(x))]^{-\frac{1}{\rho}+1-\frac{\gamma}{\rho}} v(s(x))}{\rho\, u(s(x)) + s(x)u'(s(x))} \tag{2.4.10}$$

and $s(x)$ is determined by

$$s^{\rho}u(s) = x^{\rho}, \quad \left.\frac{dx}{ds}\right|_{s=0} > 0. \tag{2.4.11}$$

In any particular instance the above reductions and transformations are tedious to carry out. It is best in such circumstances to proceed formally. The above then provides a rigorous basis for the following formal recipe.

Consider

$$\mathscr{I} = \int_{\alpha}^{\beta} e^{ith(s)} g(s)ds \tag{2.4.12}$$

with $h(s)$ monotonic in the interval.

We insert the neutralizers $\nu_{\alpha}, \nu_{\beta}$, $\nu_{\alpha}+\nu_{\beta} = 1$ and write

$$\begin{aligned} \mathscr{I} &= \int_{\alpha}^{\beta} e^{ith(s)} \nu_{\alpha}(s) g(s)ds + \int_{\alpha}^{\beta} e^{ith(s)} \nu_{\beta} g(s)ds \\ &= e^{ith_{\alpha}} \int_{\alpha}^{\beta} e^{it(h(s)-h_{\beta})} \nu_{\beta} g(s)ds \\ &\quad + e^{ith_{\beta}} \int_{\alpha}^{\beta} e^{it(h(s)-h_{\alpha})} \nu_{\alpha} g(s)ds \end{aligned}$$

we expand in the neighborhood of the two critical points

$$h(s) - h_\alpha = (s-\alpha)^\rho[u^0_\alpha + u^1_\alpha(s-\alpha) + \dots]$$

$$h(s) - h_\beta = (\beta-s)^\sigma[u^0_\beta + u^1_\beta(\beta-s) + \dots]$$

$$g(s) = (s-\alpha)^\mu[v^0_\alpha + v^1_\alpha(s-\alpha) + \dots]$$

$$g(s) = (\beta-s)^\nu[v^0_\beta + v^1_\beta(\beta-s) + \dots]$$

The power series coefficients are a consequence of the assumed differentiability of the coefficient functions u and v. Let us introduce the formal expansions,

$$\mathcal{J} = e^{ith_\alpha}\int_\alpha^\beta e^{it(s-\alpha)^\rho u^o_\alpha}\nu_\alpha[1+iu^1_\alpha(s-\alpha)^{\rho+1}t+\dots](s-\alpha)^\mu[v^0_\alpha+\dots]ds$$

(2.4.13)

$$+ e^{ith_\beta}\int_\alpha^\beta e^{it(\beta-s)^\sigma u^o_\beta}\nu_\beta[1+iu^1_\beta(\beta-s)^{\sigma+1}t+\dots](\beta-s)^\nu[v^0_\beta+\dots]ds$$

Next set $s-\alpha = x$ in the first integral and $\beta-s = x$ in the second integral to obtain

$$\mathcal{J} = e^{ith_\alpha}\int_0^\infty e^{itx^\rho u^o_\alpha}\tilde{\nu}[1+iu^1_\alpha x^{\rho+1}t+\dots][v^0_\alpha+\dots]x^\mu dx$$

$$+ e^{ith_\beta}\int_0^\infty e^{itx^\sigma u^o_\beta}\tilde{\nu}[1+iu^1_\beta x^{\sigma+1}t+\dots][v^0_\beta+\dots]x^\nu dx$$

where we take $\tilde{\nu}(x)$ to be a C^∞_0 function identically equal to one on a sufficiently large interval including the origin.

Finally setting $s = x^\rho$ and $s = x^\sigma$ as the new variable s of integration in the first and second terms respectively we obtain,

$$\mathcal{J} = \frac{e^{ith_\alpha}}{\rho}\int_0^\infty e^{itu^0_\alpha s}\tilde{\nu}[1+iu^1_\alpha s^{\frac{1}{\rho}+1}t+\dots][v^0_\alpha+\dots]s^{\frac{\mu+1-\rho}{\rho}}ds$$

(2.4.14)

$$+ \frac{e^{ith_\beta}}{\sigma}\int_0^\infty e^{itu^0_\beta s}\tilde{\nu}[1+iu^1_\beta s^{\frac{1}{\sigma}+1}t+\dots][v^0_\beta+\dots]s^{\frac{\nu+1-\sigma}{\sigma}}ds$$

then using the corollary to Theorem 215 all terms may be evaluated

$$\mathscr{I} = \frac{e^{ith_\alpha} v_\alpha^o (\frac{\mu+1-\rho}{\rho})!}{\rho(-itu_\alpha^o)^{(\mu+1)/\rho}}[1+\ldots] + \frac{e^{ith_\beta} v_\beta^o (\frac{\nu+1-\sigma}{\sigma})!}{\sigma(-itu_\beta^o)^{(\sigma+1)/\sigma}}[1+\ldots] \tag{2.4.15}$$

Subsequent orders are found by carrying further terms in the expansions.

Exercise 33. Find two terms in the AD of

(a) $\int_0^1 \cos(xt^\rho)dt$

(b) $\int_0^{\pi/2} (1-\frac{2\theta}{\pi})^{\frac{1}{2}} \cos(x\cos\theta)d\theta$

as $|x| \to \infty$, x real.

Exercise 34. Find three terms in the AD of

(a) $\int_0^1 \frac{e^{ist^{5/2}}}{t^{1/3}}dt$

(b) $\int_0^\pi e^{ist^2\cos^2 t}dt$

for $|s| \to \infty$, s real.

2.5. Integrals of the Type $\int_{\alpha(\underset{\sim}{x})}^{\beta(\underset{\sim}{x})} G(\underset{\sim}{x},t)dt$, $\underset{\sim}{x} \to \infty$.

The integrals one encounters in practice are not always in a form in which the Laplace or Kelvin or saddle point formulas apply. Generally speaking one finds integrals in the form

$$\int_{a(\underset{\sim}{x})}^{b(\underset{\sim}{x})} G(\underset{\sim}{x},t)dt \tag{2.5.1}$$

where $\underset{\sim}{x} = (x_1,\ldots,x_n)$ is an n-vector which is tending to infinity. Many difficulties appear in the asymptotic treatment of integrals of such general type. Instead of trying to anticipate all these in a rigorous development we now discuss formal procedures for obtaining the leading term in the AD of (2.5.1). Our discussion will be based on the analysis of sections 2.3 and 2.4.

Generalized Laplace Formula

Since the properties of $a(\underset{\sim}{x})$ $b(\underset{\sim}{x})$ as well as $G(\underset{\sim}{x},t)$ depend on the manner in which $\underset{\sim}{x} \to \infty$, it will be useful to make this explicit. We will write

$$\underset{\sim}{x} \to \infty, \quad \underset{\sim}{x} \in \mathscr{C}_\infty$$

to indicate that the vector $\underset{\sim}{x}$ tends to ∞ along some specific curve $\mathscr{C}_\infty$.

Next for $\underset{\sim}{x} \to \infty$, $\underset{\sim}{x} \in \mathscr{C}_\infty$ we decompose the interval $a(\underset{\sim}{x})$, $b(\underset{\sim}{x})$ into subintervals in which $G(\underset{\sim}{x},t)$ is of one sign and also in which $G(x,t)$ has a single maximum point. Assuming e.g. that $G(\underset{\sim}{x},t) > 0$ in the subinterval, which we denote by $(\alpha(\underset{\sim}{x}),\beta(\underset{\sim}{x}))$, we write

$$\ln G = h$$

On the basis of these remarks we can consider

$$R(\underset{\sim}{x}) = \int_{\alpha(\underset{\sim}{x})}^{\beta(\underset{\sim}{x})} e^{h(\underset{\sim}{x},t)}\,dt \tag{2.5.2}$$

for $\underset{\sim}{x} \to \infty$, $\underset{\sim}{x} \in \mathscr{C}_\infty$ under the assumption that in this limit $h(\underset{\sim}{x},t)$ has a single maximum point (which is also a global maximum). Also we assume that $h(\underset{\sim}{x},t)$ is continuous in $\underset{\sim}{x}$ and twice continuously differentiable in t. We denote the location of the maximum point by $t_0(\underset{\sim}{x})$ so that

$$\frac{\partial h}{\partial t}(\underset{\sim}{x},t_0(\underset{\sim}{x})) = 0 \tag{2.5.3}$$

and we also assume that

$$\frac{\partial^2 h^o}{\partial t^2} \neq 0 \; . \tag{2.5.4}$$

The zero superscript denotes evaluation at $t = t^o(\underset{\sim}{x})$. It follows from the implicit function theorem that $t^o(\underset{\sim}{x})$ is a continuous function.

Based on the discussion given in sections 2.3 and 2.4 it is plausible that the leading term in the AD of (2.5.2) is gotten by expanding $h(\underset{\sim}{x},t)$ in the neighborhood of $t_o(\underset{\sim}{x})$,

$$h = h^o + \frac{h^o_{tt}}{2}(t-t_o)^2 + h^o_{ttt}(t-t_o)^3/6 + \ldots$$

$$= h^o + h^o_{tt}(t-t_o)^2/2 + r(\underset{\sim}{x},t)$$

Then (2.5.2) takes the form

$$R(t) = e^{h^o}\int_{\alpha(\underset{\sim}{x})}^{\beta(\underset{\sim}{x})} \exp[h^o_{tt}\frac{(t-t_o)^2}{2} + \ldots]dt$$

On regarding the analysis leading to the Laplace formula, we might by analogy suppose

$$R \sim e^{h^o}\int_{\alpha(\underset{\sim}{x})}^{\beta(\underset{\sim}{x})} \exp[h^o_{tt}\frac{(t-t_o)^2}{2}]dt \tag{2.5.5}$$

Formally we would expect this to be valid if the remainder term $r(\underset{\sim}{x},t)$ is small compared to $h^o_{tt}(t-t_o)^2$. Continuing in this vein, we take the next term in the Taylor series, i.e. the third order term, as being representative of the remainder. Therefore we might expect (2.5.5) to be valid if

$$|h^o_{tt}(t-t_o)^2| >> |h^o_{ttt}(t-t_o)^3| \tag{2.5.6}$$

As it stands this criteria is vacuous since no condition has been set on the magnitude of $|t-to|$.

Now two possibilities present themselves. 1) If $|h^o_{tt}| < \infty$ for $\underset{\sim}{x} \to \infty$, $\underset{\sim}{x} \in \mathscr{L}_\infty$ then (2.5.6) suggests the criteria

$$1 >> \left| \frac{h^o_{ttt}(t-t_o)}{h^o_{tt}} \right|, \quad \text{for} \quad t \in (\alpha,\beta) \tag{2.5.7}$$

for the validity of (2.5.5). 2) On the other hand if

$$|h^o_{tt}| \to \infty, \quad \underset{\sim}{x} \to \infty, \; \underset{\sim}{x} \in \mathscr{C}_\infty$$

the major contribution to the evaluation of (2.5.5) comes from

$$(t-t_o) = O\left(\frac{1}{(-h^o_{tt})^{\frac{1}{2}}}\right) \tag{2.5.8}$$

and hence this suggests the condition

$$\frac{|h^o_{ttt}|}{|h^o_{tt}|^{3/2}} = o(1), \quad \underset{\sim}{x} \to \infty, \; \underset{\sim}{x} \in \mathscr{C}_\infty \tag{2.5.9}$$

for the validity of (2.5.5). In this case it is assumed that the interval defined by (2.5.8) is contained in (α,β).

This second possibility corresponds to an asymptotic peaking as was the case for the Laplace formula. The first situation however is a new situation and has no analogue in the Laplace formula.

To summarize we can write

Case 1. $\quad -h^o_{tt} < \infty$

$$R \sim e^{h^o} \int_{\alpha(\underset{\sim}{x})}^{\beta(\underset{\sim}{x})} \exp\left[h^o_{tt}\frac{(t-t_o)^2}{2}\right] dt$$

(2.5.10)

$$h_{ttt}(\underset{\sim}{x},t) = o(h^o_{tt}) \quad \underset{\sim}{x} \to \infty, \; \underset{\sim}{x} \in \mathscr{C}_\infty$$

Case 2. $\quad -h^o_{tt} \to \infty$

$$R \sim e^{h^o} \left(\frac{2\pi}{-h^o_{tt}}\right)^{\frac{1}{2}} \tag{2.5.11}$$

$$h^o_{ttt} = o(|h^o_{tt}|^{3/2}), \quad x \to \infty, \; x \in \mathscr{C}_\infty$$

In Case 2 the criteria $h^{o}_{ttt} = o(|h^{o}_{tt}|^{3/2})$ itself suggests the possible curves $\mathscr{C}_{\infty}$ along which the Laplace formula (2.5.11) is valid. Note also that if $h = xh(t) + \ln g(t)$ we obtain Laplace's formula (2.3.2) and the criteria in (2.5.11) is simply $x^{-1/2} = o(1)$.

<u>Example</u>. Consider

$$J(x) = \int_{0}^{\infty} \frac{e^{xt}}{t^{t}}\, dt$$

for $x \to \infty$. Following the above procedure we exponentiate t^{-t} to obtain

$$J(x) = \int_{0}^{\infty} e^{xt-t \ln t} dt$$

Considering $h(x,t) = xt - t \ln t$ we have

$$h_{t}(x,t) = x-1- \ln t$$

$$h_{tt}(x,t) = -\frac{1}{t}$$

$$h_{ttt}(x,t) = +\frac{1}{t^{2}}$$

Therefore we have a single stationary point at

$$t = e^{x-1}.$$

Although this already yields to the above procedure it is a matter of convenience to place the maximum point in the finite region. Therefore we write

$$t = e^{x-1}y$$

and we find

$$J(x) = e^{x-1}\int_0^\infty \exp[-e^{x-1}(y \ln y - y)]dy$$

which is of the form discussed in section 3. In any case we have,

$$J(x) \sim e^{x-1}\exp[e^{x-1}]\left(\frac{2\pi}{\exp(x-1)}\right)^{\frac{1}{2}}$$

$$\sim e^{x/2}\exp(e^{x})(2\pi)^{\frac{1}{2}}$$

Generalized Kelvin Formula

In a similar manner we can discuss

(2.5.12) $$K = \int_{\alpha(\underset{\sim}{x})}^{\beta(\underset{\sim}{x})} e^{ih(\underset{\sim}{x},t)}dt$$

where α,β, and h are real and $h(\underset{\sim}{x},t)$ has a single stationary point in the interval, at $t_o = t_o(\underset{\sim}{x})$; say,

$$\frac{\partial h}{\partial t}(\underset{\sim}{x},t)\Big|_{t=t_o} = 0 = h_t^o$$

Also we take

$$h_{tt}^o \neq 0$$

Then following arguments similar to those given above we are led to,

$$K \sim \left(\frac{2\pi}{|h_{tt}^o|}\right)^{\frac{1}{2}}\exp\left[ih^o + \frac{i\pi}{4}\frac{h_{tt}^o}{|h_{tt}^o|}\right]$$

(2.5.13) when

$$h_{tt}^o \to \infty, \quad \underset{\sim}{x} \to \infty, \quad \underset{\sim}{x} \in \mathscr{C}_\infty$$

and

$$h_{ttt}^o = o(|h_{tt}^o|^{3/2})$$

This then generalizes Kelvin's formula. It should be noted that we do not now consider an analogue to (2.5.10). This is due to the fact that Kelvin's formula really depends on the rapid oscillation of the integrand.

As an illustration consider

$$\mathcal{J}(t) = \int_0^\infty e^{ixt - \frac{ix^q}{q}}\,dx \tag{2.5.14}$$

$q > 1$, and $t \to \infty$. The convergence of this integral is guaranteed by the second term of the exponent. This becomes apparent with the change of variable $x^q = s$.

To formally evaluate (2.5.14) we first search for the stationary point

$$\frac{\partial}{\partial x}\left[xt - \frac{x^q}{q}\right] = 0$$

or

$$t = x^{q-1}$$

$$x_o = t^{\frac{1}{q-1}}$$

Changing variables

$$x = t^{\frac{1}{q-1}}\,s$$

(2.5.14) becomes

$$\mathcal{J}(t) = t^{\frac{1}{q-1}} \int_0^\infty e^{\left[+it^{\frac{q}{q-1}}\cdot\left(s - \frac{s^q}{q}\right)\right]}\,ds$$

The problem then simply reduces to a Kelvin type integral and applying (2.4.15) we find,

$$\mathcal{J}(t) \sim \left(\frac{2\pi}{q-1}\right)^{\frac{1}{2}} \cdot e^{\left[i\left(\frac{q-1}{q}\right)t^{q/(q-1)} - i\frac{\pi}{4}\right]} \cdot \frac{1}{t^{(q-2)/2(q-1)}}$$

Dispersive Wave Propagation

As a second illustration we consider an important class of integrals, which occur in problems of wave propagation,

$$W(x,t) = \int_{-\infty}^{\infty} e^{i[\omega(k)t - kx]}\phi(k)dk \tag{2.5.15}$$

The "frequency" $\omega(k)$ is a real function of the "wave number" k. In keeping with our discussion we seek the AD of (2.5.15) when

$$(x,t) \to \infty$$

Proceeding formally we search for a stationary point of the exponent in (2.5.15)

$$t\frac{\partial\omega(k)}{\partial k} - x = 0 \tag{2.5.16}$$

Write $\xi = \frac{x}{t}$ and let one such root of (2.5.16) be represented by $k_o(\xi)$, i.e.

$$\frac{\partial\omega(k_o)}{\partial k_o} = \xi$$

[The quantity $\frac{\partial\omega}{\partial k}$ is referred to as the group velocity and in problems of physical origin it represents the speed at which energy is transmitted. In this connection the ratio ω/k is referred to as the phase or signal speed]. We therefore write

$$W(x,t) \sim e^{i[\omega(k_o)t-k_ox]}\phi(k_o)\int_{-\infty}^{\infty} e^{i\omega_o''\frac{s^2t}{2}}ds$$

And evaluating the integral

$$W(x,t) \sim e^{i[\omega(k_o)t-k_ox]}\phi(k_o)e^{\frac{i\pi}{4}\frac{\omega_o''}{|\omega_o''|}}\left(\frac{2\pi}{t|\omega_o''|}\right)^{\frac{1}{2}} \tag{2.5.17}$$

Which of course resembles Kelvin's formula. The formal condition for the validity of (2.5.15) is

$$\frac{|\omega_0'''|}{\sqrt{t}\,|\omega_0''|^{3/2}} << 1 \tag{2.5.18}$$

It is sometimes useful to employ the condition (2.5.18) to introduce a new large parameter, i.e. we write

$$s = \frac{\sqrt{t}\,|\omega_0''|^{3/2}}{|\omega_0'''|} \tag{2.5.19}$$

and then rewrite the exponent of (2.5.15) as

$$\omega(k)t - kt = sf(k;\ x,t)$$

and seek an evaluation for $s \to \infty$.

To illustrate these notions we consider an example from water wave theory. For small amplitude shallow water waves one finds that the frequency is $\omega = (g|k|)^{\frac{1}{2}}$ or

$$F(x,t) = \int_{-\infty}^{\infty} e^{i((g|k|)^{\frac{1}{2}} t-kx)} F(k)dk \tag{2.5.20}$$

where g is the gravitational constant. This type of phenomena may be made more familiar by noting that the signal speed is proportional to the square root of the wave length of a wave - a fact which emerges through a simple dimensional argument.

By a simple change of variable (2.5.20) becomes

$$\begin{aligned} F(x,t) &= \int_0^{\infty} e^{i(t(gk)^{\frac{1}{2}} - kx)} F(k)dk \\ &\quad + \int_0^{\infty} e^{i(t(gk)^{\frac{1}{2}} + kx)} F(-k)dk \\ &= F^+(x,t) + F^-(x,t) \end{aligned}$$

It is clear that the first term represents a superposition of waves traveling to the right and F^- a superposition of waves traveling to the left. It will suffice to consider just the first term

(2.5.21)
$$F^+ = \int_0^\infty e^{i((gkt)^{\frac{1}{2}} - kx)} F(k)dk$$

Condition (2.5.16) for the location of the stationary point yields

(2.5.22)
$$k_o = \frac{gt^2}{4x^2}$$

The formal consistency condition now is

(2.5.23)
$$s = \frac{gt^2}{x} >> 1$$

Following the above suggestion of taking (2.5.23) as the new large parameter, we set

$$k = \frac{sp}{x} = \frac{gt^2}{x^2} p$$

Under this change of variable (2.5.21) becomes,

$$F^+ = \int_0^\infty e^{is(\sqrt{p} - p)} F(\frac{sp}{x}) dp \cdot \frac{s}{x}$$

We can now use Kelvin's formula directly, and find

$$F^+ \sim e^{\frac{igt^2}{4x} - \frac{i\pi}{4}} (\frac{gt^2 \pi}{x^3})^{\frac{1}{2}} \cdot F(\frac{gt^2}{4x^2})$$

and

$$t >> \sqrt{x}$$

[Many of the formal procedures of this section have been given a rigorous foundation in W. C. Huo and L. Sirovich, "Stationary Exponent Formulas for Many Large Parameters", D.A.M. Report #32, Brown University.]

Exercise 35. Find the leading term in the AD of

$$\int_0^\infty x^{-\alpha x} t^x dx$$

for $\alpha > 0$, $t \to \infty$

Exercise 36. For what values of α and β real do the following integrals converge?

$$\int_0^\infty e^{-xt-x^\alpha t^\beta} dx, \quad \int_0^\infty e^{xt-x^\alpha t^\beta} dx, \int_0^\infty e^{-xt+x^\alpha t^\beta} dx$$

Find the leading term in the AD of each as $t \to \infty$.

Exercise 37. Find the leading term in the AD of

$$\int_1^\infty e^{i(x - \frac{t}{x})} dx$$

for $t \to \infty$.

2.6. Method of Steepest Descents and The Saddle Point Formulas

We now consider integrals of the form

$$I(t) = \int_a^b e^{tf(x)} g(x) dx$$

again in the limit $t \to \infty$, but with f a complex valued function

$$f = \phi + i\psi.$$

Except for the coincidental case when ψ is stationary at the global maximum of ϕ the previous techniques now do not apply. In order to develope a theory of sufficient scope, it will be necessary to require that f and g be complex analytic functions and we henceforth write

$$I(t) = \int_{T_{ab}} g(z)\exp[tf(z)]dz \tag{2.6.1}$$

T_{ab} signifies a contour in the complex z-plane between the endpoints a and b, and entirely contained in an open region R, in which f and g are both analytic.

By Cauchy's theorem we may use instead of T_{ab} any other path T'_{ab} in R into which T_{ab} may be continuously distorted. It is this property which we wish to exploit. As motivation, note that

$$|I(t)| \leq \int_{T_{ab}} e^{t\phi}|g(z)dz|$$

and denoting the maximum of ϕ on T_{ab} by ϕ_m

$$|I(t)| < e^{t\phi_m}\int_{T_{ab}} |g(z)dz|$$

Then assuredly we will obtain our best estimate by choosing a possible T_{ab} on which ϕ_m is minimum (and for which $\int_{T_{ab}} |g\,dz|$ is bounded).

Before following this up we introduce certain standard definitions.

Definition. A path T on which $\mathrm{Re}f = \phi$ is constant, is said to be a level curve of f.

Definition. $\mathrm{Im}f = \psi$ is called the phase of f.

Definition. A path T, which is a path of stationary phase of a function f(z), will also be referred to as a path of steepest incline of f(z) or simply a steepest path of f.

The basis for the first and second definitions is clear. As motivation for the last definition consider the directional derivative in the θ-direction given by

$$\cos\theta \frac{\partial}{\partial x} + \sin\theta \frac{\partial}{\partial y} = \frac{\partial}{\partial s}$$

where s represents arc length. The necessary condition that ϕ increase or decrease most rapidly in a direction θ is simply,

$$\frac{\partial}{\partial \theta} \phi = 0$$

Reverting to cartesian derivatives the condition is

$$\frac{\partial \phi}{\partial \theta} = \frac{\partial x}{\partial \theta}\frac{\partial}{\partial x} \phi + \frac{\partial y}{\partial \theta}\frac{\partial}{\partial y} \phi = 0$$

$$= -\sin\theta \frac{\partial}{\partial x}\phi + \cos\theta \frac{\partial}{\partial y}\phi$$

And on using the Cauchy-Riemann equations, the condition is

$$\frac{\partial \psi}{\partial s} = 0.$$

If on a path of steepest incline $\frac{\partial \phi}{\partial s} \lesseqgtr 0$ we shall say it is a path of steepest $\left\{\begin{matrix}\text{descent}\\ \text{ascent}\end{matrix}\right\}$ of f.

Returning to (2.6.1) we for the sake of definiteness assume

$$\phi(z=a) > \phi(z=b) \tag{2.6.2}$$

We now distinguish two cases

Case I. There exists a path T'_{ab} belonging to R so that $\phi(a) > \phi(z)$, $z \in T_{ab}$, $z \neq a$.

Case II. Otherwise

The two situations are sketched in the following two figures.

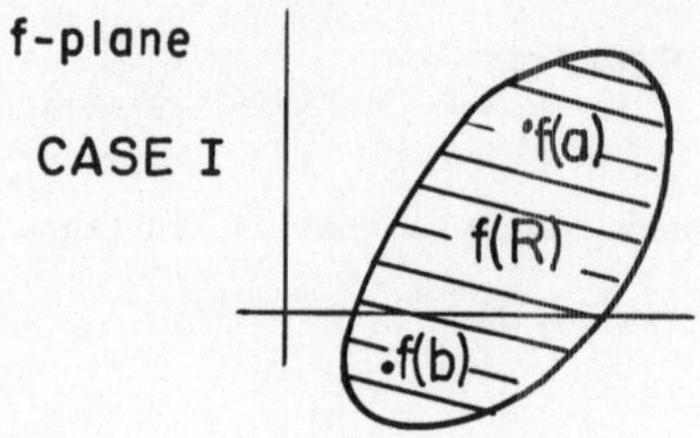

Figure 1

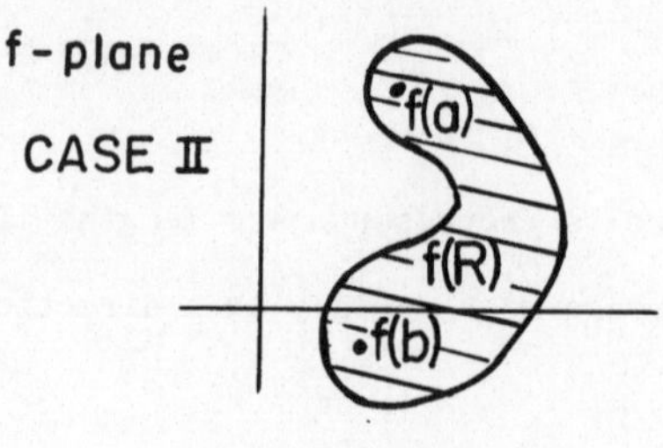

Figure 2

The hatched region in each represents the image under $f(z)$ of R, i.e. $f(R)$. In the first case it is clear that the image curves $f(T)$ can be distorted so that $\phi(a) > \phi(z)$, $z \in T$, $z \neq a$. For the second figure it is equally clear that this cannot be done.

<u>Case I.</u>

Since $f(R)$ is an open set we can find a T^*_{ab}, $f(T^*_{ab}) \in f(R)$ so that there is a finite segment leading away from $f(a)$ which is a path of steepest descent. $f(T^*_{ab})$ is sketched in the figure

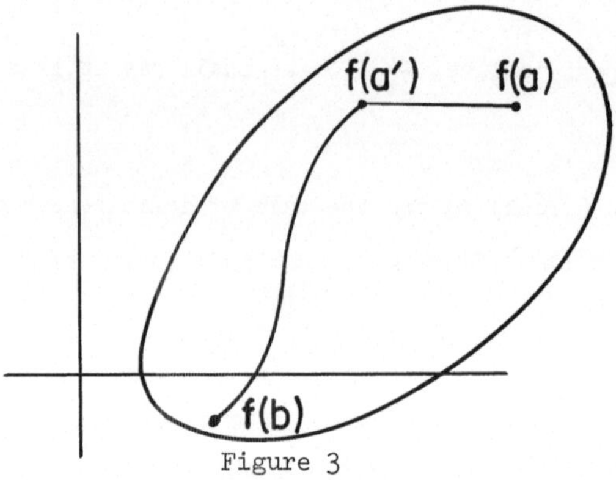

Figure 3

Along the portion of T^*_{ab} going from $a \to a'$ $\text{Im } f = \psi = \psi(a)$. On the path $a' \to b$ of T^*_{ab} $\text{Re} f = \phi \leq \phi(a') < \phi(a)$.

Therefore we may write

$$(2.6.3)\qquad \begin{aligned} I(t) &= \int_{T_a} e^{tf(z)} g(z)dz + \int_{T^*_{ab}-T_a} e^{tf(z)} g(z)dz \\ &= I_o + \hat{I} \end{aligned}$$

T_a represents the path from $a \to a'$ along which $\operatorname{Im} f = \psi(a)$. Since any precise value of a' will be immaterial we ignore the dependence of T_a on a'. From the above remarks we may estimate $\hat{I}(t)$ by

$$\hat{I}(t) = O(\exp[(a')t])$$

Also we can write I_o as

$$(2.6.4)\qquad I_o = e^{f(a)t} \int_{T_a} g(z)\exp[(\phi(z)-\phi(a))t]dz$$

The integral I_o is therefore a Laplace integral and can be handled by the methods of section 2.4. In this sense we have concluded the discussion of this case. Actually we will be more specific about the evaluation of the integral of (2.6.4), at the end of this section.

In the above we have assumed that $\phi(a) > \phi(b)$. If $\phi(b) > \phi(a)$ the same type of discussion applies. If $\phi(a) = \phi(b)$ we find two integrals of the type (2.6.4). In certain cases, singularities of the function $g(z)$ may prevent the distortion of T_{ab} into T^*_{ab}. This is not a serious difficulty as is illustrated by the following exercise.

Exercise 38. Find the leading term in the AD of

$$I(t) = \int_{T_{-1,+1}} e^{-izt} \frac{dz}{\sqrt{z}}$$

where $T_{-1,1}$ is a semi-circle in the low half plane.

Case II. Along T_{ab}, $\phi_m = \max_{z\in T_{ab}} \phi(z)$, $\phi_m \neq \phi(a), \phi(b)$ for all possible $T_{ab} \in T$.

Let ϕ^*_m be the minimum such value and which is taken on at $z^* \in R$ (we are now assuming that this is an interior point). There may be a number of such points but for simplicity we assume a single one. Let T^*_{ab} be a path containing z^* and let γ^* be a path of steepest incline through z^*. Denote arclength along γ^* by ℓ and along T^*_{ab} by s. Then

$$\left.\frac{\partial\phi}{\partial\ell}\right|_{z=z^*} = 0$$

for otherwise there would be a neighboring value of $\phi < \phi^*_n$. Also since $\phi(z^*) = \phi^m$ for T^*_{ab}

$$\left.\frac{\partial\phi}{\partial s}\right|_{z=z^*} = 0$$

But since $\phi_{,\ell} = \phi_{,s} = 0$ it follows that $\phi_{,x} = \phi_{,y} = 0$ and from the Cauchy-Riemann equations

$$\left.\frac{\partial f}{\partial z}\right|_{z=z^*} = 0$$

To go further we must analyze the situation when there exists a $z^* \in R$ such $f'(z^*) = 0$. Again for simplicity we assume that there is only one such point. Also for simplicity we take

$$c = f''(z^*) \neq 0 \tag{2.6.5}$$

These restrictions will be easily lifted later in our discussion.

A great simplification in the discussion is effected by the canonical transformation

$$f(z) - f^* = \frac{c}{2}\zeta^2 \tag{2.6.6}$$

Then it follows from function theory that there are two one-to-one analytic branches $z(\zeta)$. To fix the branch we take $z(\zeta)$ such that

(2.6.7) $$\left.\frac{dz}{d\zeta}\right|_{z*} = 1$$

Under (2.6.6), the $I(f)$ becomes

$$I(t) = e^{f*t} \int_{T_{\alpha\beta}} e^{\frac{c\zeta^2}{2}} \chi(\zeta) d\zeta$$

where

$$\chi(\zeta) = g(z(\zeta))\frac{dz}{d\zeta}$$

is analytic and $T_{\alpha\beta}$ is the image of T_{ab},

$$\alpha = \zeta(a), \quad \beta = \zeta(b)$$

Next we write

$$F = \Phi + i\Psi = \frac{c}{2} \zeta^2$$

and set

$$-\theta_0 = \frac{1}{2} \arg f''(z*), \quad \theta = \arg \zeta$$

Then $\Psi = 0$ for

$$\theta = \theta_0, \quad \theta_0 + \frac{\pi}{2}, \quad \theta_0 + \pi, \quad \theta_0 + \frac{3\pi}{4}$$

which are paths of steep incline and $\Phi = 0$ for

$$\theta = \theta_o + \frac{\pi}{4}, \quad \theta_o + \frac{3\pi}{4}, \quad \theta_o + \frac{5\pi}{4}, \quad \theta_o + \frac{7\pi}{2}$$

which are level lines. These as well as other Ψ = constant (steepest paths), Φ = constant (level curves) are plotted in the figure 4.

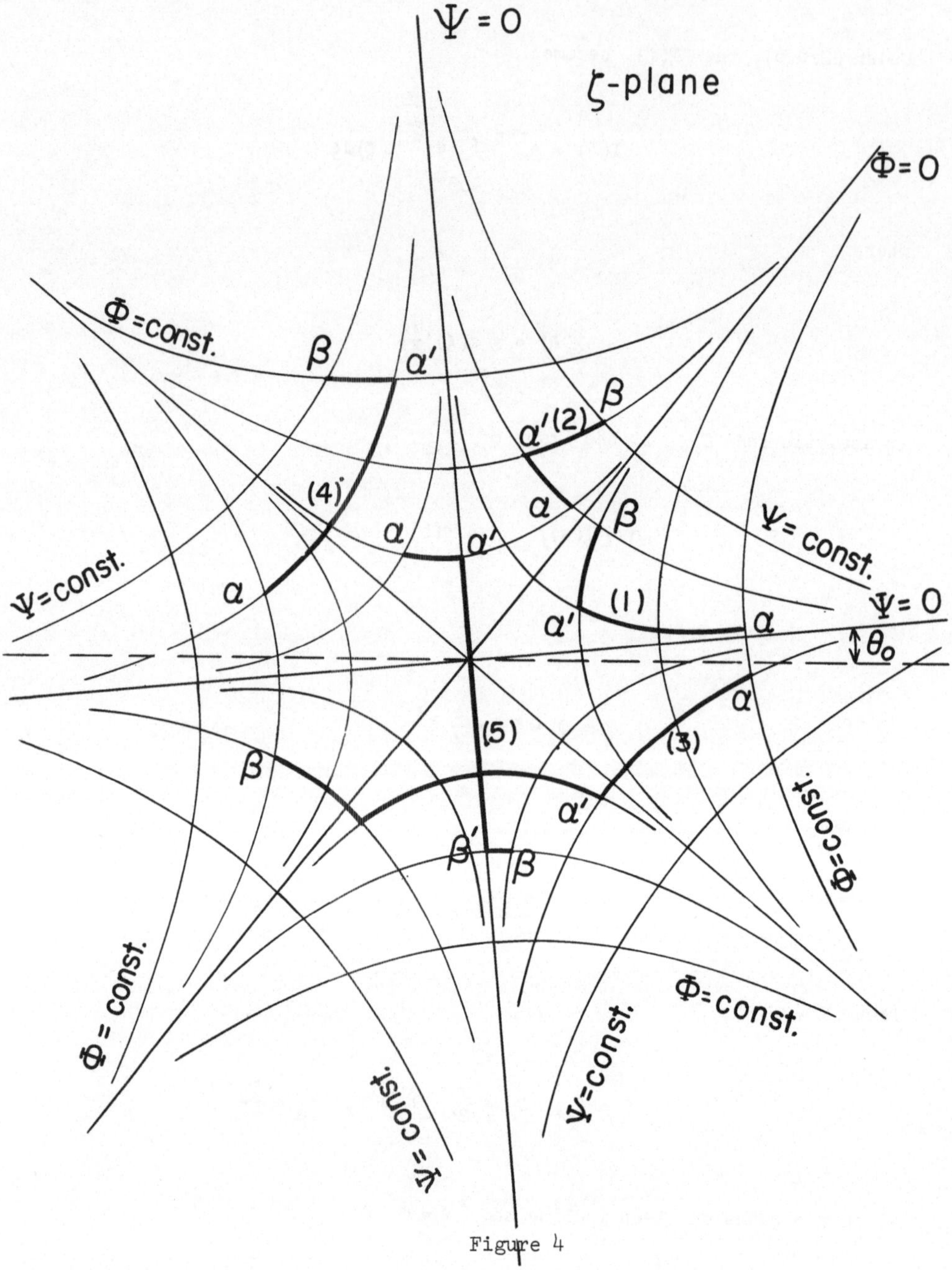

Figure 4

Regarding figure 4 we see that the level lines $\Phi = 0$ divide the plane into hills and valleys as indicated on the sketch below

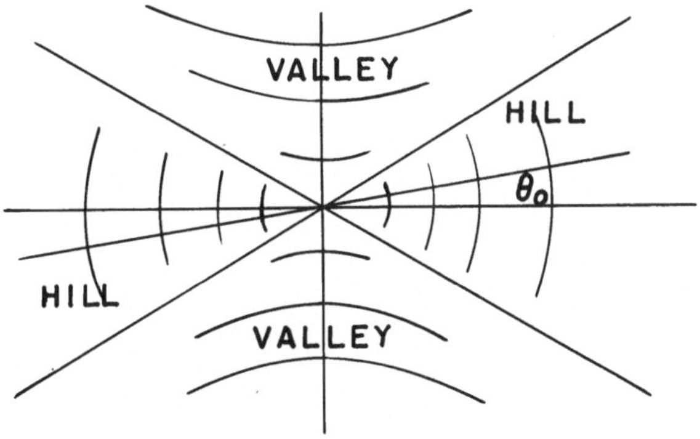

and for obvious reasons z^* is referred to as a saddle point (Note that a similar sketch for Ψ appears also as a saddle, rotated by $\pi/4$) There are now five possible situations in regard to the location of the endpoints α and β (or for that matter a and b).

(1) The endpoints on same hill

(2) The endpoints in same valley

(3) The endpoints on different hills

(4) One endpoint in a valley and one on a hill

(5) Both endpoints in different valleys.

Examining figure 4 we see that situations (1) - (4) correspond to the discussion in Case I. On taking the paths indicated by (1) $\rightarrow$ (4) we obtain images of possible T^*_{ab} with each of the paths $\alpha - \alpha'$ being the images of T_a ($\alpha = \zeta(a)$, $\alpha' = \zeta(a')$).

Also examining figure 4 we see how to deal with situation (5). For on passing from one valley to another by means of the steepest path $\Psi = 0$, we achieve the same situation discussed in Case I, i.e. we write

$$(2.6.8) \qquad I = \int_{T_{z^*b(\beta)}} e^{f(z)t} g(z)dz - \int_{T_{z^*a(\alpha)}} e^{f(z)t} g(z)dz$$

Next we distort $T_{z^*b(\beta)}$ and $T_{z^*a(\alpha)}$ into paths whose combined image is (5) of figure 4. We can therefore write

$$I = -I_{oa} + O(e^{\phi(a)t}) + I_{ob} + O(e^{\phi(b)t})$$

with

$$(2.6.9) \qquad I_{oa} = \int_{T_{z^*(a)}} e^{f(z)t} g(z)dz$$

and

$$(2.6.10) \qquad I_{ob} = \int_{T_{z^*(b)}} e^{f(z)t} g(z)dz$$

$T_{z^*(a)}$ is the path of steepest descent from z^* into the valley of a and $T_{z^*(b)}$ the path of steepest descent from z^* into the valley of b. As before we can also write

$$(2.6.11) \qquad I_{oa,b} = e^{f^*t} \int_{T_{z^*(a,b)}} g(z) \exp[(\phi(z)-\phi(z^*))t]dz$$

which is in the form of a Laplace integral.

We have assumed that $f''(z^*) \neq 0$, however it is generally to be expected under Case II that

$$(2.6.12) \qquad f(z) = f^* + \frac{(z-z^*)^m}{m!} f^{(m)}(z^*) + \ldots$$

with $m \geq z$. It is clear from this that in general we have m hills and m

valleys, e.g. when $m = 3$ we have the sketch

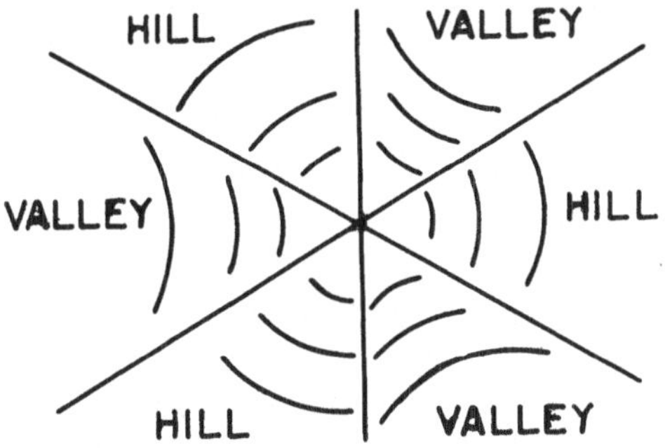

(sometimes called a monkey saddle). The same five situations listed in the previous paragraph still exhaust all the possibilities and the analysis of the general case $m \geq 2$ is not really different than for $m = 2$ discussed above.

By essentially geometrical considerations we have reduced the AD of (2.6.1) to the AD of a Laplace integral, and the asymptotic expansion follows from the considerations in section 2.3. This being the case it is tempting to unload the restriction of t to real values and in certain instances in the literature this suggestion is made. It is however in general incorrect. We now demonstrate this and at the same time develope the analysis for complex t - which in fact is very simple.

Saddle Point Method for a Complex Large Parameter

We write

$$t = |t| e^{i\Lambda} = s e^{i\Lambda} \tag{2.6.13}$$

and consider the integral

$$I(t) = \int_{T_{z^*b}} e^{tf(z)} g(z) dz \tag{2.6.14}$$

This is regarded as representative of (2.6.9,10) or (2.6.4) say. The point z^* has the property that $\phi^* > \phi(z)$, $z \neq z^*$, $z \in T_{z^*b}$. It may or may not be a saddle.

A typical sketch in the f-plane is shown below.

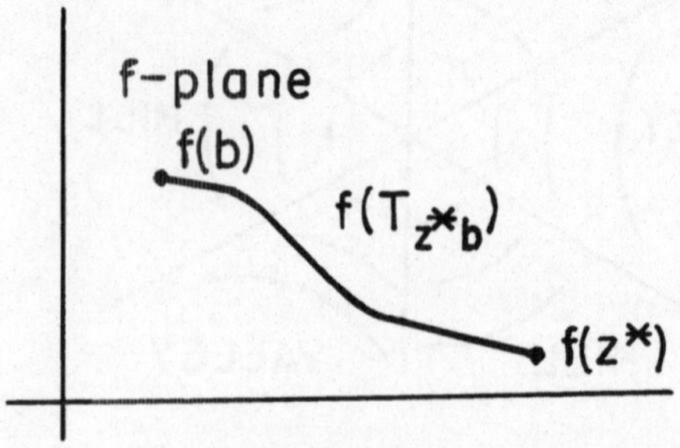

Figure 5

From the maximum property of φ^* we can say the

(2.6.15) $$|\theta| = |\arg[f^* - f(b)]| < \frac{\pi}{2}$$

Next inserting (2.6.13) into (2.6.14) the case of complex z is formally the same as real t if we write

$$I(t) = \int e^{s\hat{f}(z)} g(z)dz$$

where

$$\hat{f} = e^{i\Lambda}f$$

Clearly the asymptotic development will still only depend on the point z^* when

(2.6.16) $$|\arg[\hat{f}^* - \hat{f}(b)]| = |\theta + \Lambda| < \frac{\pi}{2}$$

(We are here assuming that the straight line between f^* and $f(b)$ lies in the common region of analyticity.) Moreover the rays

(2.6.17) $$\Lambda = \pm \frac{\pi}{2} - \theta$$

are Stokes lines for the function $I(t)$, (2.6.11). It is also clear that in the ranges

$$\frac{3\pi}{2} - \theta > \Lambda > \frac{\pi}{2} - \theta$$

(2.6.18)

$$-\frac{3\pi}{2} - \theta < \Lambda < -\frac{\pi}{2} - \theta$$

the main contribution comes from the $z = b$ endpoint, and that $\frac{3\pi}{2} \pm \theta$ are Stokes lines and so forth. In the event that we are dealing with a saddle point situation we must consider both the integrals of (2.6.8) to find the correct AD of (2.6.1) as Λ changes. For in this case it is clear that changes in Λ can alter a situation (5) to situations (3) or (4).

Exercise 39. Consider the entire function,

$$\int_{-1}^{2} e^{ws^2} ds$$

for all $\arg w$, $|w| \to \infty$. Specify the Stoke's lines and the lead term in the AD for all sectors.

The Complete Asymptotic Development

Having made the above remarks we now regard t as real and positive. Since certain special features enter in the asymptotic calculation of $I_0(t)$ we consider it in some detail. For (2.6.4) or (2.6.9,10) we consider

(2.6.14) $$I_0(t) = \int_{T_{z*}} e^{tf(z)} g(z)dz$$

where the path of integration is determined by the condition

(2.6.20) $$\text{Im } f = \text{Im } f*$$

and is a curve of steepest descent of $f(z)$. The final endpoint is of course of no importance in the AD. In the neighborhood of z^* we have (2.6.12) which we now write as

$$f = f^* + f^{(m)*}\frac{(z-z^*)^m}{m!} + \dots = f^* - (z-z^*)^m h(z)$$

with $h(z^*) \neq 0$. If $m > 1$ we say that z^* is saddle point of order m.

Next we set

$$(z-z^*)^m h(z) = \tau \tag{2.6.21}$$

Regarding τ as a complex variable for the moment (recalled that τ is real on the path of integration), analytic function theory tells us that there exist m roots of (2.6.21) such that

$$z = z(\tau)$$
$$z^* = z(0)$$

Further if we write $s = \tau^{1/m}$, z is analytic in s, i.e. there exists a convergent power series expansion,

$$z = z^* + c_1 s + c_2 s^2 + \dots$$

Also if ω_m denotes a primative m-th root of unity (so that ω_m^k, $k = 1,2,\dots,m$ give all the roots of unity) the other branches have the expansion

$$z = z^* + c_1(s\omega_m^k) + c_2(s\omega_m^k)^2 + \dots, \quad k = 1,2,\dots,m \tag{2.6.22}$$

From (2.6.20) we know we can take $\tau > 0$, and we can write

(2.6.23) $$I_o(t) = -e^{f^*t} \int_0 e^{-\tau t} \frac{g(z(\tau))}{f'(z(\tau))} d\tau$$

Specifying the upper limit precisely is immaterial since it does not contribute in the asymptotic evaluation, it of course is positive. In order to now use Watson's lemma we need the expansion of

(2.6.24) $$\chi(\tau) = -\frac{g(z(\tau))}{f'(z(\tau))}$$

at $\tau = 0$. If $m = 1$ this is straight forward since χ is analytic in τ. If $m > 1$ some care is necessary. For in this case it is necessary to take the branch of $z(\tau)$ so that as τ increases $z(\tau)$ enters the valley of a or b as the case may be.

To carry out the expansion of $\chi(\tau)$ we first consider (2.6.21) in the form

$$\tau = -\frac{f^{(m)}(z^*)(z-z^*)^m}{m!} - \cdots$$

Then substituting (2.6.20) we find

(2.6.25) $$z = z^* + \left[\frac{-m!}{f^{(m)}(z^*)}\right]^{1/m} \tau^{1/m} \omega_m^k + O(\tau^{2/m})$$

where we take $\tau^{1/m} > 0$, i.e. we take the principal branch, and $[-m!/f^{(m)}(z^*)]^{1/m}$ be any fixed root of the quantity in the brackets. We now choose the exponent k in the ω_m^k so that z enters the a-valley when τ increases and denote this by $\omega_m(a)$ and similarly $\omega_m(b)$ represents the m-th root of unity that sends (2.6.25) into the b-valley. The expansion of $\chi(\tau)$ is perhaps best carried out by again introducing s but now through

(2.6.26) $$s = \omega_m \tau^{1/m}$$

when ω_m is either $\omega_m(a)$ or $\omega_m(b)$. Then we write

$$\chi = g(z(s))\frac{dz}{ds}\frac{ds}{d\tau} = g(z(s))\frac{dz}{ds}\frac{\omega_m \tau^{\frac{1}{m}-1}}{m}$$

The expansion of $g(z(s))\frac{dz}{ds}$ is relatively simple since both $g(z(s))$ and $z'(s)$ are analytic in s. [The only sensitive point lies in choosing the branch of $z(s)$ (there are m) so that it combined with (2.6.6) and yields (2.6.25).] Hence we can write

$$g(z(s))\frac{dz}{ds} = \sum_{n=0} a_n s^n = \sum_{n=0} a_n \omega_m^n \tau^{n/m} \tag{2.6.27}$$

For example

$$a_0 = g^*[-m!/f^{(m)}(z^*)]^{1/m} \tag{2.6.28}$$

Substitution into (2.6.23) then yields

$$I_0(t) \sim e^{tf^*}\frac{\omega_m}{m}\sum_{n=0}\frac{a_n\omega_m^n(\frac{n+1-m}{m})!}{t^{(n+1)/m}}$$

Or returning to the original integral $I(t)$, in the case of a single saddle and situation (5),

$$I \sim -I_{oa} + I_{ob}$$

from (2.6.9) and

$$I(t) \sim \frac{e^{tf^*}}{m}\cdot\sum_{n=0}^{\infty}\frac{a_n(\frac{n+1-m}{m})!}{t^{(n+1)/m}}[\omega_m^{n+1}(b) - \omega_m^{n+1}(a)] \tag{2.6.29}$$

Using (2.6.28) the lead term is

(2.6.30) $$I(t) \sim \frac{e^{tf^*} g^*\Gamma(\frac{1}{m})}{mt^{1/m}}[\omega_m(b) - \omega_m(a)][-m!/f^{(m)}(z^*)]^{1/m}$$

An alternate expression is gotten by writing

$$f^{(m)}(z^*) = -Re^{-i\theta^*}$$

and noting that the m-roots of unity can be expressed as

$$e^{2\pi ik/m}, \quad k = 0,1,\dots,m-1$$

Then we can write (2.6.30) as

(2.6.31) $$I(t) \sim \frac{g^*\Gamma(\frac{1}{m})}{mt^{1/m}}\exp[tf^*][\frac{m!}{R}]^{1/m}\{\exp[\frac{i(\theta^*+2\pi k_b)}{m}] - \exp[\frac{i(\theta^*+2\pi k_a)}{m}]\}$$

The condition on the integer $k_a(k_b)$ is that

$$\exp[\frac{i(\theta^*+2\pi k_a)}{m}](\exp[\frac{i(\theta^*+2\pi k_b)}{m}])$$

enter the a(b) valley.

When $m = 2$ (2.6.31) can be written as

$$I(t) \sim \frac{e^{tf^*} g^*\Gamma(\frac{1}{2})}{2\sqrt{t}}[e^{ik_b\pi} - e^{i(k_b+1)\pi}][\frac{-2!}{f''(z^*)}]^{1/2}$$

(2.6.32) or

$$I(t) \sim e^{tf^*} g^*[\frac{2\pi}{-tf''(z^*)}]^{\frac{1}{2}}$$

with the radical such that $\arg(\frac{2\pi}{-tf''(z^*)})^{\frac{1}{2}}$ is in the b-valley. (2.6.32) is sometimes referred to as the saddle point formula.

Exercise 40. Find lead term of AD of

(a) $\int_0^{\pi i} e^{te^2} \frac{dz}{\sqrt{z}}$

(b) $\int_{1-\frac{i}{3}}^{1+\frac{i}{3}} e^{tz^2} \frac{dz}{1+z}$

as $t \to \infty$.

Exercise 41. Find lead term of AD, as $t \to \infty$

(a) $\int_1^i e^{tz^3} dz$

(b) $\int_{-i}^1 e^{tz^4} dz$

Application to Bessel Functions

As an illustration of the method of steepest descents, consider the Bessel function (Magnus, Oberhettinger, and Soni [14], p. 82)

$$H_0^{(1)}(w) = \frac{1}{\pi i} \int_{-\infty}^{\infty+i\pi} e^{w \sin hz} dz$$

for $|w| \to \infty$ and for the moment we can regard w as real positive. The path of integration is as sketched below

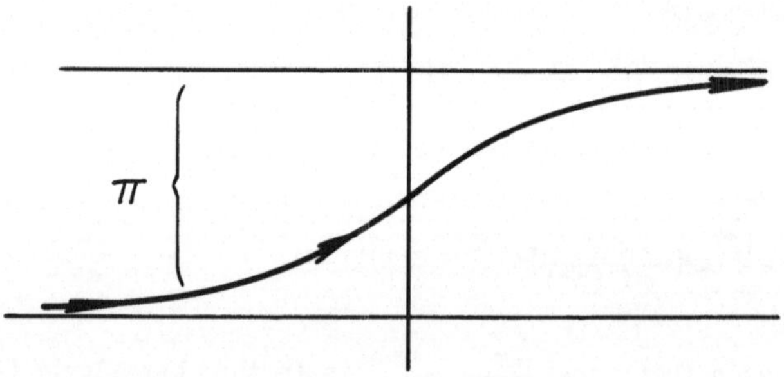

To locate the saddle,

$$\frac{d}{dz} \sin h\, z' = 0 = e^z + e^{-z}$$

or

$$e^{2z} = 1$$

and therefore

$$z = \frac{\pi i}{2} + in\pi, \quad n = 0, \pm 1, \ldots$$

Since it seems likely that the pertinent saddle is $\frac{\pi i}{2}$, we transform variables to

$$\eta = z - \frac{\pi i}{2}$$

Then since $\sinh z = \sinh(\eta + \frac{\pi i}{2}) = i \cosh \eta$, we obtain

(2.6.33) $$H_0^{(1)}(w) = \frac{1}{\pi i} \int_{-\infty-\frac{i\pi}{2}}^{\infty+\frac{i\pi}{2}} e^{wf(\eta)} d\eta \; e^{iw} = \frac{2e^{iw}}{i\pi} \int_0^{\infty+\frac{i\pi}{2}} e^{wf(\eta)} d\eta$$

with

(2.6.34) $$f(\eta) = (\cosh \eta - 1)i$$

and

$$f(0) = 0 = f'(0)$$
$$f''(0) = i$$

The level lines given by

$$\mathrm{Re}(i \cosh \eta - i) = 0$$

or

$$\text{Re}[i\{e^{x}(\cos y + i \sin y)\} + e^{-x}\{\cos y - i \sin y\} - i] = 0$$

or

$$\sin y \sinh x = 0.$$

Hence the level lines are

$$y = n\pi$$
$$x = 0$$

And similarly the steep paths $\text{Im } f(\eta) = 0$ are

$$(e^{x} + e^{-x})\cos y - 2 = 0$$

A sketch of the terrain is given below,

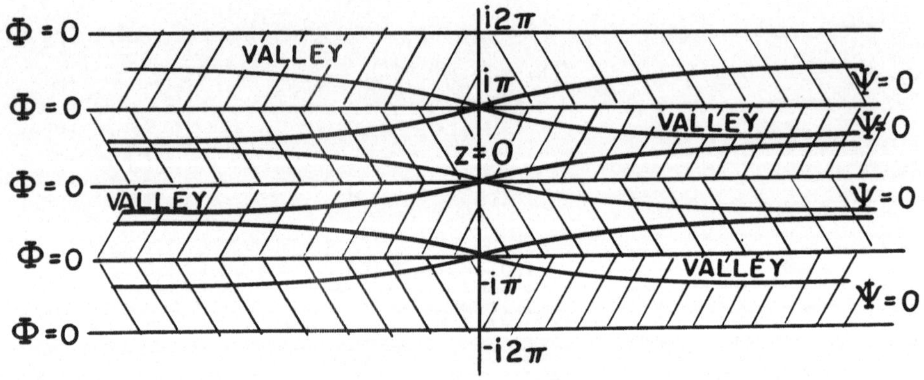

The heavily marked path is the appropriate one for our integration.

Actually the above geometrical discussion is mostly unnecessary. We only require the knowledge that the original contour of integration may be distorted to go from valley to valley over a saddle.

Setting

$$-\tau = i \cosh \eta - i$$

in (2.6.33) we obtain

$$H_0^{(1)} = \frac{2e^{iw}}{\pi i} \int_0^\infty e^{-w\tau} \frac{d\eta}{d\tau} d\tau$$

where the integration is along the real line. Next consider

$$\frac{d\eta}{d\tau} = \frac{i}{\sinh \eta} = \frac{i\omega}{(\cosh^2 \eta - 1)^{\frac{1}{2}}} = \frac{\omega}{(\tau^2 - 2i\tau)^{\frac{1}{2}}}$$

where ω is a square root of unity and hence is ± 1. Regarding the radical as the principal branch $((\tau^2-2i\tau)^{\frac{1}{2}} \to \tau, \quad \tau \to \infty)$ it is clear that $\omega = +1$. Therefore

(2.6.35) $$H_0^{(1)}(w) = \frac{2e^{iw}}{\pi i} \int_0^\infty \frac{e^{-w\tau}}{(\tau^2-2i\tau)^{\frac{1}{2}}} d\tau$$

It is important to note in (2.6.35) we have equality and therefore the expansion

$$\frac{1}{(\tau^2-2i\tau)^{\frac{1}{2}}} = \frac{e^{-\frac{i\pi}{4}}}{(2\tau)^{\frac{1}{2}}} [1 - \frac{i\tau}{4} \pm \ldots].$$

is valid for

(2.6.36) $$-\frac{3\pi}{2} < \arg \tau < \frac{\pi}{2}$$

Hence

(2.6.37) $$H_o^{(1)}(w) \sim \frac{2e^{iw}}{\pi i} \cdot \frac{e^{-\frac{i\pi}{4}}}{\sqrt{2}}\left[\frac{\Gamma(\frac{1}{2})}{w^{1/2}} - \frac{i(\frac{1}{2})!}{w^{3/2}} \pm \ldots\right]$$

is valid for $|w| \to \infty$ and

$$-\pi < \arg w < 2\pi$$

from (2.6.36) and Generalized Watson's Lemma, Theorem 124.

Exercise 42. A representation of the Bessel function $J_\nu(w)$ is (Magnus, Oberhettinger and Soni [14], p. 82)

(2.6.38) $$J_\nu(z) = \frac{1}{2\pi i} \int_{\infty - i\pi}^{\infty + i\pi} \exp[w \sinh t - \nu t]dt$$

Find the lead term of the AD of (2.6.38).

2.7. Applications of the Saddle Point Method

The Airy Integral

We first consider the Airy functions. These functions play a central role in the discussion of uniform expansions in section 2.11 and also in the WKB approximation, section 3.7, and hence merit a discussion in their own right.

We consider the Airy functions defined by

(2.7.1) $$A(z) = \frac{1}{2\pi i} \int_{\mathcal{L}} \exp[tz - \frac{t^3}{3}]\, dt$$

with the path of integration $\mathcal{L}$: $\mathcal{L}_o$, $\mathcal{L}_+$, $\mathcal{L}_-$ is indicated in the figure

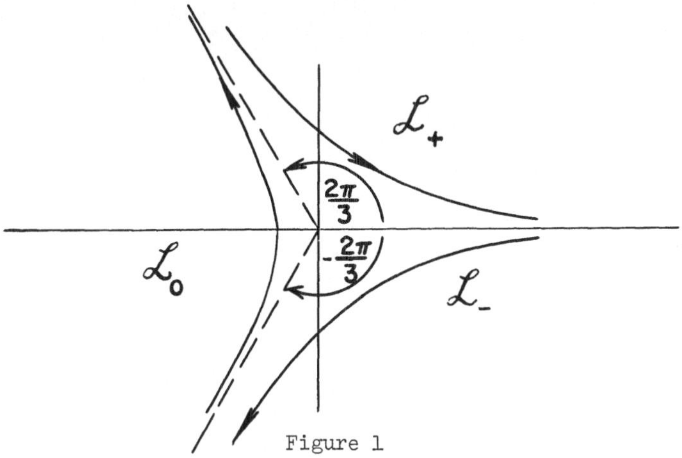

Figure 1

Note that t^3 is real positive on each of the asymptotes. Equation (2.7.1) denotes the generic form of the Airy functions, and we will write A_0, $A_\pm$ for the specific forms.

Each of the Airy function A are entire and clearly

(2.7.2) $$A_0 + A_+ + A_- = 0.$$

Also it may be verified by substitution that

(2.7.3) $$\frac{d^2A(z)}{dz^2} - zA(z) = 0$$

which is the Airy equation.

As a point of note we mention that the two independent solutions of (2.7.3) are generally written as $Ai(z)$ and $Bi(z)$. In our notation

$$Ai(z) = A_0(z)$$
$$Bi(z) = iA_+(z) - iA_-(z)$$

Also the relationship to Bessel functions of one-third order should be noted

$$A_o(z) = \frac{1}{\pi}\left(\frac{z}{3}\right)^{\frac{1}{2}} K_{\frac{1}{3}}\left(\frac{2}{3} z^{3/2}\right)$$

(See [15], sec. 3.4 and [16], sec. 10.4).

Under the changes of variable

$$te^{\pm 2\pi i/3} = t'$$

in (2.7.1), we easily see

$$A_+(z) = e^{\mp 2\pi i/3} A_{\underset{-}{o}}(ze^{\mp 2\pi i/3})$$

(2.7.4) $$A_o(z) = e^{\mp 2\pi i/3} A_{\underset{+}{-}}(ze^{\mp 2\pi i/3})$$

$$A_-(z) = e^{\mp 2\pi i/3} A_{\underset{o}{+}}(ze^{\mp 2\pi i/3})$$

Hence a knowledge of the functions $A(z)$ in a sector of angle $2\pi/3$ determines the functions everywhere. Alternately a knowledge of any of the Ai in the entire plane entirely determines the remaining functions.

We now focus attention on $A_o(z)$. Introducing the variable transformation

$$t = z^{1/2}\tau,$$

where we choose the principal branch of the square root, we obtain

$$A_o = \frac{z^{1/2}}{2\pi i}\int_{\arg(z^{1/2})\mathcal{L}_o} \exp[z^{3/2}(\tau-\tau^3/3)]d\tau$$

Taking $\arg z$ to be sufficiently small ($|\arg z^{1/2}| < \pi/6$) we can deform the path of integration $\arg(z^{1/2})\mathcal{L}_o$ back into $\mathcal{L}_o$, so that

(2.7.5) $$A_o(z) = \frac{z^{1/2}}{2\pi i}\int_{\mathcal{L}_o} \exp[z^{3/2}(\tau-\tau^3/3)]d\tau$$

By analytic continuation this defines $A_0(z)$ for all z.

In order to apply the saddle point analysis we consider

$$f(\tau) = \tau - \tau^3/3 = \varphi + i\psi$$

This has the saddle points

$$f'(\tau = \pm 1) = 0$$

and in neighborhood of these points

$$f(\tau) \sim \pm 2/3 \mp (\tau \mp 1)^2 - \frac{1}{3}(\tau \mp 1)^3$$

Note that the saddle at $\tau = 1$ is higher than that at $\tau = -1$. A sketch of the terrain is given in the following figure.

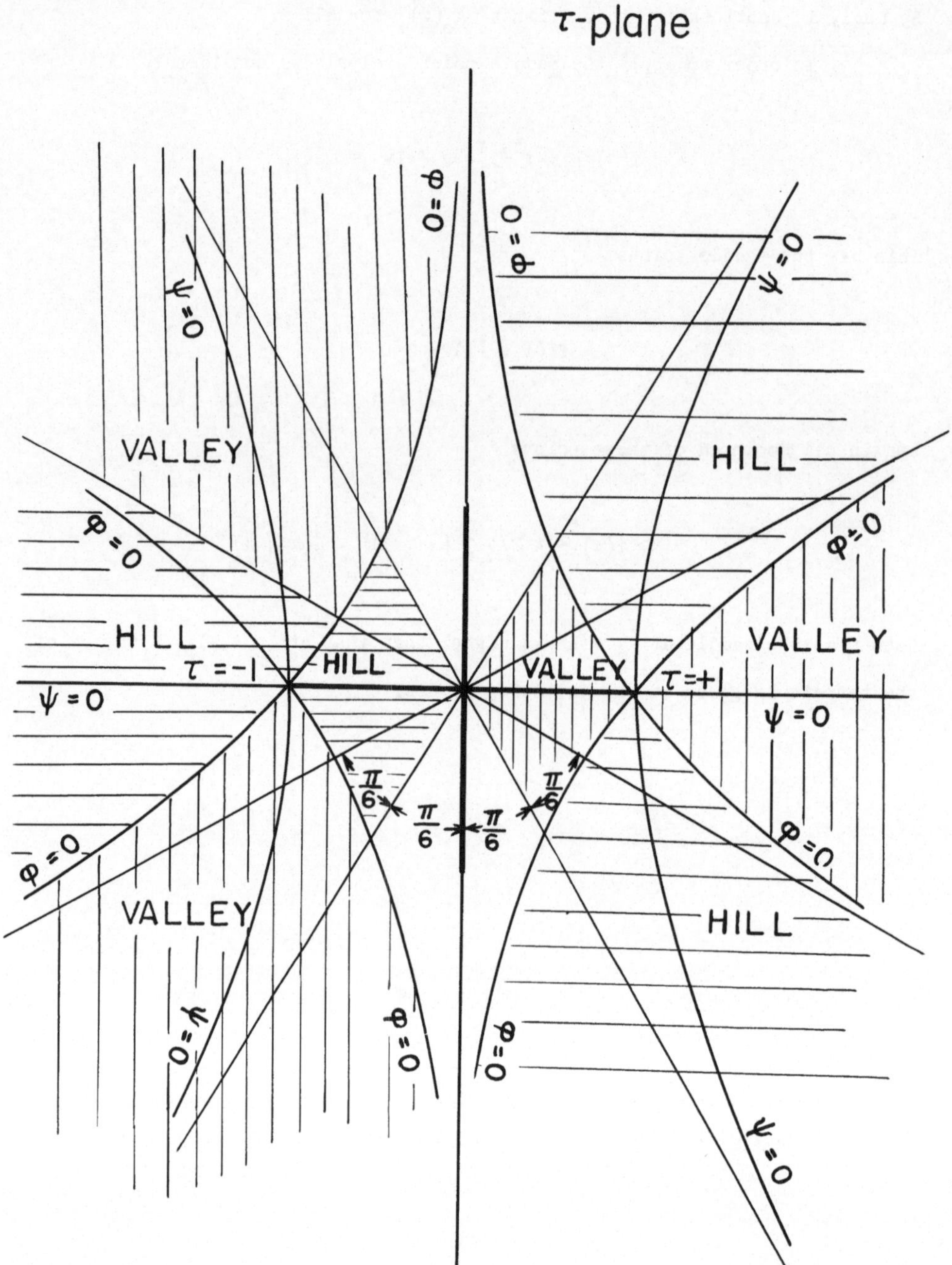
τ-plane
φ=0
ψ=0
VALLEY
HILL
HILL
τ=-1
HILL
ψ=0
VALLEY
VALLEY
τ=+1
ψ=0
π/6
π/6
π/6
π/6
VALLEY
HILL

Figure 2

It is clear from figures 1 and 2 that $\mathcal{L}_0$ may be taken as the path of steepest descents passing through $\tau = -1$. This done we consider contributions from the third and fourth quadrants separately. We write

$$A_0 = \frac{z^{1/2}}{2\pi i}\left(\int_{\mathcal{L}_0^+} - \int_{\mathcal{L}_0^-}\right)\exp[z^{3/2}(\tau - \frac{\tau^3}{3})]d\tau$$

where $\mathcal{L}_0^+(\mathcal{L}_0^-)$ is the part of $\mathcal{L}_0$ going from $\tau= -1$ into the third (fourth) quadrant. Next we make the variable transformation

$$s = \frac{\tau^3}{3} - \tau - \frac{2}{3} = \frac{(\tau-2)}{3}(\tau+1)^2 \tag{2.7.6}$$

$$\frac{d\tau}{ds} = \frac{1}{\tau^2-1} \tag{2.7.7}$$

$\tau(s)$ has two branches corresponding to $\mathcal{L}_0^+$ and $\mathcal{L}_0^-$, these we denote by τ^+ and τ^-, hence

$$A_0(z) = \frac{z^{1/2}e^{-\frac{2}{3}z^{3/2}}}{2\pi i}\int_0^\infty e^{-sz^{3/2}}\left(\frac{d\tau^+}{ds} - \frac{d\tau^-}{ds}\right)d\tau \tag{2.7.8}$$

and the path of integration is the positive axis (since $\tau^3/3 - \tau - \frac{2}{3}$ varies between $0,\infty$ on $\bar{\mathcal{L}}_0$).

The integral in (2.7.8) is in the form of a Laplace integral and hence for z large we need the expansion of $\frac{d\tau}{ds}$ in the neighborhood of $s = 0$, and hence $\tau = -1$. In the neighborhood of $\tau = -1$ it is clear that τ has an expansion of the form

$$\tau = -1 + \alpha_1 s^{1/2} + \alpha_2 s + \alpha_3 s^{3/2} + \ldots$$

Substituting into (2.7.6) we find to lowest order

$$\alpha_1^2 = -1$$

Since $\tau^+(\tau^-)$ must enter the third (fourth) quadrant for increasing s we must have

$$\alpha_1^{\pm} = \pm i$$

From this it follows that e.g.,

$$\alpha_2^{\pm} = -1/6, \quad \alpha_3^{\pm} = \mp 5i/72$$

so that

$$\tau^{\pm} = -1 \pm is^{1/2} - s/6 \mp 5is^{3/2}/72 + \ldots \tag{2.7.9}$$

and

$$\left(\frac{d\tau^+}{ds} - \frac{d\tau^-}{ds}\right) = 2i/s^{1/2} - 15is^{1/2}/72 + \ldots \tag{2.7.10}$$

In order to determine the sectors in which (2.7.9) and (2.7.10) are valid, we note from (2.7.7) that $\frac{d\tau}{ds}$ (and hence $\tau(s)$) is singular for $\tau = \pm 1$ and therefore for $s = 0, -4/3$. Therefore (2.7.9,10) hold for

$$|\arg s| < \pi \tag{2.7.11}$$

Next substituting (2.7.10) into (2.7.8) and using Watson's lemma, we find

$$A_o(z) \sim \frac{\exp[-\frac{2}{3}z^{3/2}]}{2\sqrt{\pi}\, z^{1/4}} \sum_{k=0} \frac{(-)^k \Gamma(3k+1/2)}{6^{2k} k!\,\Gamma(k+1/2) z^{3k/2}} \tag{2.7.12}$$

where we have supplied the result of finding all the coefficients.

From (2.7.11) and the Generalized Watson's lemma, theorem 124, we have that (2.7.12) is valid for

$$|\arg z^{3/2}| < \frac{3\pi}{2}$$

and hence

(2.7.13) $$-\pi < \arg z < \pi.$$

So far we have ignored the placing of the branch cut for $z^{1/4}$ (and hence for $z^{1/2}$). It is convenient to place it along the Stoke's line, (2.7.13).

Considering just the first term of our result we can schematically represent our result as

(2.7.14) $$A_o(z) \sim \frac{\exp[-\frac{2}{3} z^{3/2}]}{z^{1/4}(4\pi)^{\frac{1}{2}}}$$

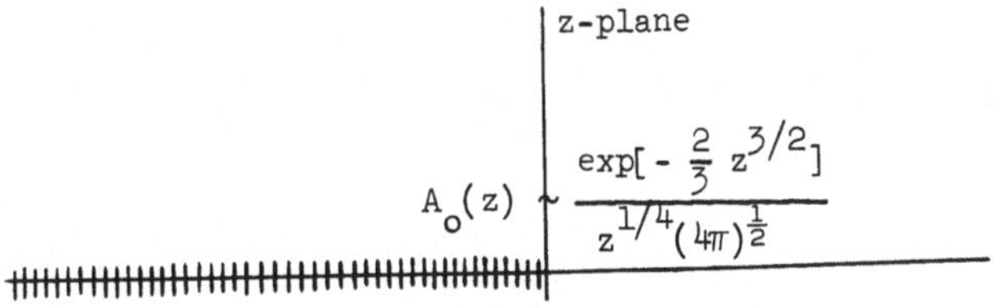

Figure 3

The asymptotic expansions of A_+ and A_- follow from (2.7.4). In particular corresponding to Figure 3 we obtain

(2.7.15) $$A_+ \sim \frac{e^{\frac{2}{3} z^{3/2}}}{i\text{-}z^{1/4}(4\pi)^{\frac{1}{2}}}$$

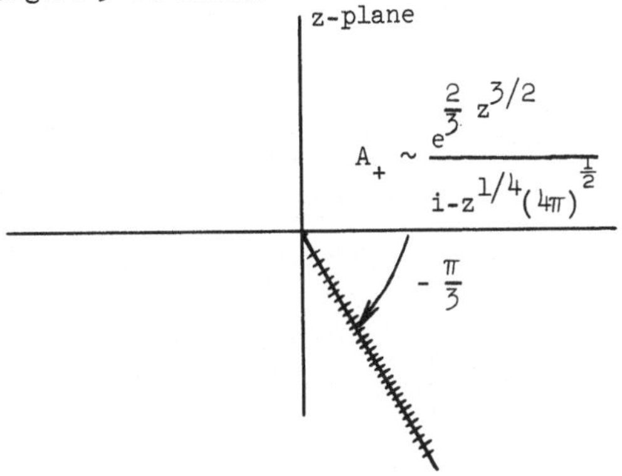

Figure 4

and

$$Ai_{-} \sim -\frac{e^{\frac{2}{3}z^{3/2}}}{iz^{1/4}(4\pi)^{\frac{1}{2}}} \tag{2.7.16}$$

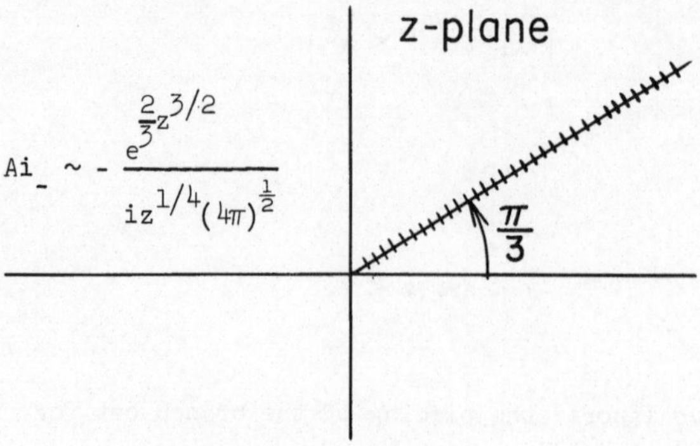

Figure 5

In each we adhere to the convention that the branch cut lies along the Stokes line.

Finally we observe that the asymptotic expansion on the Stokes lines follow from (2.7.2). In particular for A_o we have to lowest order

$$A_o \sim \frac{1}{\sqrt{\pi}\sqrt{x}} \cos\left(\frac{2}{3}x^{2/3} - \frac{\pi}{4}\right) \tag{2.7.17}$$

$$\text{for} \quad z = -x, \quad x > 0$$

From (2.7.2) one may compute any desired number of terms of the AD of each $Ai(z)$ along its Stokes line. This we leave as an exercise.

<u>A Generalization of the Airy Integral</u>

Based on various properties different generalizations of the Airy functions are to be found in the literature. In this vein we consider solutions of

$$\frac{d^n P}{dz^n} - zP = 0 \tag{2.7.18}$$

which are satisfied by the entire functions,

$$P_n(z;n) = \frac{1}{2\pi i} \int_{C_r} \exp\left[-\frac{t^{n+1}}{n+1} + zt\right] dt \tag{2.7.19}$$

with paths C_r extending from $\infty \exp[2i\pi(r+1)/n+1]$ to $\infty \exp[2i\pi\ r/n+1]$, $r = 0,1,\ldots,n$. This generates $n+1$ solutions having the relation

$$\sum_{i=0}^{n} P_i = 0$$

Setting $te^{2ip\pi/(n+1)} = t'$ in (2.7.19) we obtain the rule,

$$P_r(z) = e^{-2ip\pi/(n+1)} P_{r+p}(ze^{-2ip\pi/n+1}) \tag{2.7.20}$$

(We hence forth suppress the argument n in P_n.) Therefore a knowledge of any $P_r(z)$ in the entire plane furnishes a full description of all the P_r. (Alternately full knowledge of all the P_r in a sector of angle $2\pi i/(n+1)$ gives information in the entire plane.)

Setting $t = z^{1/n}\tau$ in (2.7.19) and for the moment taking $\arg z$ sufficiently small we obtain ($z^{1/n}$ denotes the principal branch)

$$P_r(z) = \frac{z^{1/n}}{2\pi i} \int_{C_r} \exp[-\rho(\frac{\tau^{n+1}}{n+1} - \tau)]d\tau$$

$$\rho = z^{(n+1)/n}$$

By analytic continuation this now defines $P_r(z)$ for all $\arg z$.

The exponential coefficient

$$f(\tau) = -\frac{\tau^{n+1}}{n+1} + \tau$$

has its saddle points located at the n roots of unity

$$\theta_k = e^{\frac{2\pi im}{n}}, \quad m = 0,1,2,\ldots,n-1$$

and takes on the value

$$f(\theta_k) = \frac{n\theta_k}{n+1}$$

at these points.

The remainder of the discussion now follows the treatment of Airy integral (2.7.1) (which is $P(z;2)$) and we leave it as an exercise.

2.8. Multidimensional Integrals: Part I. Laplace, Kelvin and Related Formulas

In this section we consider integrals of the form

$$R = \iint_B \psi(x,y)\exp[-z\varphi(x,y)]dxdy \tag{2.8.1}$$

in the limit $z \to \infty$, ($\mathrm{Re}\, z \geq 0$) and where B is a compact domain. (For infinite domains the neighborhood of infinity usually contributes negligibly so that this assumption is not severe.) Although a number of the calculations extend beyond two dimensions we do not attempt the general case.

We shall assume that the original domain of integration has been sufficiently sub-divided so that ψ and φ are each sufficiently smooth in each sub-domain. Also we take the bounding arcs of the subdomains, denoted by $u(x,y) = 0$, $v(x,y) = 0, \ldots$, to also be sufficiently smooth. At this point we regard the integral as having been suitably prepared in this way, so that (2.8.1) represents one of its contributions. Also to avoid further discussion of this point we take

$$\psi, \varphi \in C^\infty$$

and also the bounding arcs of B,

$$u(x,y) = 0, \quad v(x,y) = 0, \ldots \in C^\infty.$$

It was shown by J. Focke (Berichte der Sächsischen Akademie der Wissenschaften zu Leipzig, Band 101 (1954)) that the main contribution to (2.8.1) arises from the neighborhoods of certain critical points. The situation is not unlike the one dimensional integrals we have considered up to the present. We first heuristically indicate the appearance of these critical points.

To begin with suppose that $\frac{\partial\varphi}{\partial x} \neq 0$ for x,y belonging to B. Then by parts integration

$$\begin{aligned} R &= -\frac{1}{z}\int_B \frac{\psi}{\varphi_x}\frac{\partial}{\partial x} e^{-z\varphi}dxdy \\ &= \frac{1}{z}\int_B e^{-z\varphi}\frac{\partial}{\partial x}\left(\frac{\psi}{\varphi_x}\right)dxdy \\ &\quad -\frac{1}{z}\int_{\partial B}\frac{\psi}{\varphi_x}e^{-z\varphi}dy \end{aligned} \tag{2.8.2}$$

where the second integral is around the boundary B. Continuing this process indefinitely we find that R is $O(z^{-\infty})$ plus integrals along the perimeter ∂B. The latter are just one-dimensional integrals which we have investigated at length. From this we know that only endpoints and stationary points will contribute. Since φ and ψ are smooth (by construction) the endpoints arise only from discontinuities in the boundary.

<u>Definition.</u> Boundary points at which ∂B has a discontinuous derivative of some order will be called a critical point of type III.

If we denote by ℓ a variable along ∂B, then R will contribute at points where $\frac{\partial\varphi}{\partial\ell} = 0$, i.e., at stationary points. To get an invariant way of saying this note that $(u_y, -u_x)$ is in the direction of ∂B when $u(x,y) = 0$ is the boundary curve. Therefore $\frac{\partial\varphi}{\partial\ell} = 0$ when

$$(u_y, -u_x)\cdot\nabla\varphi = \begin{vmatrix}\varphi_x & \varphi_y \\ u_x & u_y\end{vmatrix} = 0 \tag{2.8.3}$$

Definition. Points of the boundary where (2.8.3) hold are called critical points of type II.

Finally we define;

Definition. A point (x,y) at which $\nabla\varphi = 0$, but at which some higher derivative is not zero, will be referred to as a critical point of type I.

We next remove the assumption that $\varphi_x \neq 0$ in B. Let us first enumerate the above discussed critical points by (x_1,y_1), $(x_2,y_2),\ldots,(x_k,y_k)$. Denote by $\alpha_i(x,y) \in C^\infty$ the "neutralizer" (see section 2.2) which is identically 1 on a sufficiently small neighborhood covering (x_i,y_i) and identically zero outside a sufficiently small neighborhood enclosing the first neighborhood. Then according to Focke

(2.8.4)
$$R = \sum_{i=1}^{k} \iint \alpha_i \psi e^{-z\varphi} dxdy + O(z^{-\infty})$$

Actually this is easy to see. Let us write

$$R = \sum_{i=1}^{k} \iint \alpha_i \psi e^{-z\varphi} dxdy + r$$

with

$$r = \iint (1 - \sum_{i=1}^{k} \alpha_i) \psi e^{-z\varphi} dxdy$$

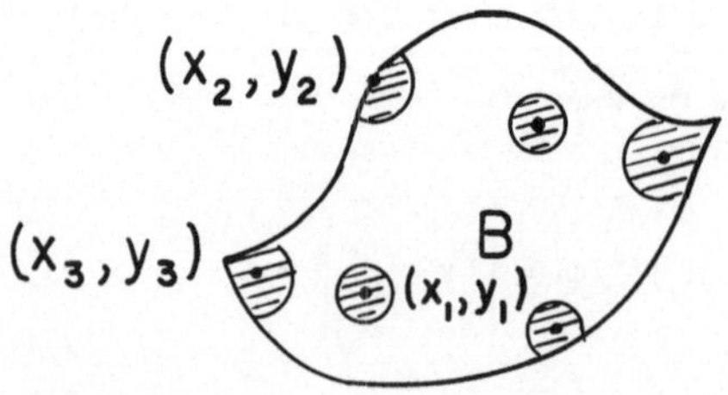

The various locations of critical points and their deleted neighborhoods are indicated in the above sketch. Denoting by $\tilde{B}$, the region B minus the deleted neighborhoods we can write

$$r = \iint_{\tilde{B}} \tilde{\psi} e^{-z\varphi} dxdy$$

with $\tilde{\psi} \in C^{\infty}$, $\nabla\varphi \neq 0$ at every point of $\tilde{B}$, and $\tilde{\psi} \equiv 0$ at "endpoints" on ∂B. Let $U_1, U_2, \ldots$ denote a sequence of neighborhoods, and corresponding functions $\beta_i(x,y) \in C^{\infty}$ such that $0 \leq \beta_i \leq 1$ for $(x,y) \in U_i$ and $\beta_i \equiv 0$ for $(x,y) \notin U$. Also such that at least one of the components of $\nabla\varphi$ is of one sign in each neighborhood U_i. And finally such that

$$\sum \beta_i = 1$$

Such decompositions of unity are standard in analysis (see Gelfand - Shilov, Generalized Functions, vol. 1) and the details play no role. We merely make use of the fact that

$$\iint_{\tilde{B}} \beta_i \tilde{\psi} e^{-z\varphi} dxdy = O(z^{-\infty})$$

which follows from repeated parts integration as in (2.8.2), except that the line integrals now vanish. (For integrals including the boundary ∂B, line integrals along ∂B do appear but these also are of $O(z^{-\infty})$ again by indefinite parts integration.)

Returning to (2.8.4) we point out that the exact form of a neutralizer $\alpha_i(x,y)$ is immaterial. To see this suppose that α and $\tilde{\alpha}$ are neutralizers for the same point then by the same arguments as used above we have

$$\iint_{B} (\alpha - \tilde{\alpha})\psi \exp(-z\varphi) dxdy = O(z^{-\infty})$$

This allows us to always choose the most convenient neutralizer in a specific calculation.

We have therefore demonstrated (2.8.4) so that it is only necessary to consider R in the neighborhoods of the critical points as indicated in the above sketch. Moreover the exact form of the neutralizer at any point is open to us.

In obtaining the asymptotic development of the terms of (2.8.4) we shall only use formal techniques. The rigorous treatment of these integrals follows the methods used in their one dimensional counterparts. A rigorous treatment is to be found in the cited paper by Focke. Also in this connection and for extensions the following papers should be consulted; D. Jones and M. Kline, J. Math. Phys. 37(1958); N. Chako, J. Inst. Maths. Applic. 1(1965); N. Bleistein and R. Handelsman, Jour. Math. Anal. Appl. 27, 2(1969).

In the sitations considered below it is sometimes necessary to distinguish between

(a) $\mathrm{Re}\, z = 0, \quad z = i\kappa$

and

(b) $|\arg z| < \frac{\pi}{2} - \varepsilon, \ \varepsilon > 0.$

In the second situation, (b), it is naturally supposed that φ is a global minimum at the critical points.

Case 1. Critical Point of Type I

We consider

$$R_I = \iint \alpha_1(x,y)\psi(x,y)\exp(-z\varphi(x,y))dxdy \tag{2.8.5}$$

where α_1 is a neutralizer (of sufficiently small support) located at the critical point (x_o,y_o) of type I. Using a zero subscript to denote evaluation at (x_o,y_o) we set

(2.8.6)
$$\underset{\sim}{\Phi}_0 = \frac{1}{2}\begin{pmatrix} \varphi_{oxx} & \varphi_{oxy} \\ \varphi_{oxy} & \varphi_{oyy} \end{pmatrix}$$

Without loss of generality we take the critical point to be located at the origin. Also it is more convenient to use vector notation so that we write $\underset{\sim}{x} = (x,y)$. In this notation we can write the Taylor expansion of φ as,

(2.8.7)
$$\varphi = \varphi_0 + (\underset{\sim}{x}, \underset{\sim}{\Phi}_o \underset{\sim}{x}) + O(x^3)$$
$$= \varphi_o + (\underset{\sim}{x}, \underset{\sim}{\Phi}_o \underset{\sim}{x}) + \varphi_r$$

and (2.8.5) as

$$R_I = \int \alpha_1(x) e^{-z(\underset{\sim}{x}, \underset{\sim}{\Phi}\, \underset{\sim}{x})} \psi(\underset{\sim}{x}) e^{-z\varphi_r} d\underset{\sim}{x}.$$

The product $\psi \exp(-z\varphi_r)$ has the expansion,

(2.8.8)
$$\psi e^{-z\varphi_r} = (\psi_0 + \underset{\sim}{x}\cdot \nabla_0 \psi_0 + \frac{(\underset{\sim}{x}\cdot \nabla_0)^2 \psi_0}{2!} + \ldots)(1 + z\,\frac{(\underset{\sim}{x}\cdot \nabla_0)^3 \varphi_0}{3!}\ \ldots)$$

(Compare this with (2.3.10)). (2.8.7,8) can now be inserted in (2.8.5) and the term by term evaluation gives the asymptotic development of R_I. The calculation is facilitated by introducing the orthogonal transformation, $\underset{\sim}{T}$, which diagonalizes $\underset{\sim}{\Phi}_0$, i.e.,

$$\underset{\sim}{T}\, \underset{\sim}{\Phi}_0 \underset{\sim}{T}' = \underset{\sim}{\Lambda}$$

the matrix of eigenvalues of $\underset{\sim}{\Phi}_0$. Then set

$$\underset{\sim}{T}\, \underset{\sim}{x} = \underset{\sim}{s} = (s_1, s_2)$$

and we have

$$R_I \sim e^{-z\varphi_0} \iint e^{-z(\underset{\sim}{s}, \underset{\sim}{\Lambda} \underset{\sim}{s})} (\psi_0 + \underset{\sim}{T}' \underset{\sim}{s} \cdot \nabla_0 \psi_0 + \dots)(1 + z \frac{(\underset{\sim}{T}' \underset{\sim}{s} \cdot \nabla_0)^3 \varphi_0}{3!} + \dots) \alpha(s_1) \alpha(s_2) ds_1 ds_2$$

We have also made use of the latitude in the choice of neutralizers by writing the neutralizer as a product of one-dimensional neutralizers. The calculation now follows that given in (2.3.10) and (2.4.13) and we go no further. It is useful to write down the leading term of the expansion, this is

(2.8.9) $$R_I \sim e^{-z\varphi_0} \psi_0 \iint e^{-z(\underset{\sim}{x}, \underset{\sim}{\Phi} \underset{\sim}{x})} dxdy = \frac{e^{-z\varphi_0} \pi}{(z^2 \det \underset{\sim}{\Phi}_0)^{1/2}}$$

or

(2.8.10) $$R_I \sim \frac{2\psi_0 \exp(-z\varphi_0)}{[z^2(\varphi_{0xx}\varphi_{0yy} - \varphi_{0xy}^2)]^{1/2}}$$

where the principal value square root is to be taken

Exercise 43. Give the details in the evaluation of the integral in (2.8.9).

Note that if $|\arg z| < \pi/2$, $\det \underset{\sim}{\Phi}_0 > 0$, while if $\operatorname{Re} z = 0$ this may take either sign.

Exercise 44 . Find the lead term of R_I when the critical point is on the boundary. Assume $|\arg z| < \pi/2$.

Exercise 45. Assuming $\det \underset{\sim}{\Phi}_0 = 0$ and $|\arg z| = \frac{\pi}{2}$ find the lead term of R_I.

Case 2. Critical Point of Type II

We consider

(2.8.11) $$R_{II} = \iint \alpha_2 \psi e^{-z\varphi} dxdy$$

where the critical point is of type II (which for simplicity we assume to be the origin) lies on the boundary curve

$$u(x,y) = 0$$

Also at the critical point we have from (2.8.3),

$$\begin{vmatrix} \varphi_{ox} & \varphi_{ox} \\ u_{ox} & u_{oy} \end{vmatrix} = 0 \tag{2.8.12}$$

The formal theory is straightforward. We assume that $u(x,y)$ increases from the boundary on entering B (otherwise we can use $-u$ instead of u). Then we introduce new coordinates

$$u = u(x,y)$$
$$s = s(x,y)$$

where the level lines of s are the orthogonal trajectories to the level lines of u. Introducing this into (2.8.11) we obtain

$$R_{II} = \iint \tilde{\alpha}(s,u)\tilde{\psi}(s,u)e^{-z\tilde{\varphi}(s,u)} \frac{\partial(x,y)}{\partial(s,u)} dsdu$$

where for example $\tilde{\varphi}(s,u) = \varphi(x(s,u), y(s,u))$ and $j = \frac{\partial(x,y)}{\partial(s,u)}$ is the jacobean. In the neighborhood of the critical point, which we fix to be the origin, we have

$$\tilde{\varphi} = \varphi_o + u\tilde{\varphi}_{ou} + \frac{s^2}{2}\tilde{\varphi}_{oss} + su\,\varphi_o su + \frac{u^2}{2}\varphi_{ouu} + \ldots$$

since it follows from (2.8.12) that $\varphi_{os} = 0$. Taking the transformed neutralizer $\tilde{\alpha}$, to be a product of one dimensional neutralizers and formally expanding we obtain

$$R_{II} \sim e^{-z\varphi_o} \int_{-\infty}^{\infty} \alpha(s)ds \int_0^{\infty} \alpha(t)dt(\psi_o+\ldots)(j_o+\ldots)(1-zsu\varphi_{osu}+\ldots) \cdot e^{-z(u\varphi_{ou}+s^2\varphi_{oss})} \tag{2.8.13}$$

The calculation now follows that given in (2.3.10) and (2.4.13) and we do not take (2.8.13) any further.

It is useful however to have the lead term in the expansion of R_{II}. To obtain this let us observe that

$$(2.8.14) \qquad \underset{\sim}{e}_u = \left(\frac{u_{ox}}{|\nabla u_o|}, \frac{u_{oy}}{|\nabla u_o|}\right)$$

is a unit vector in the u-direction and

$$(2.8.15) \qquad \underset{\sim}{e}_s = \left(-\frac{u_{oy}}{|\nabla u_o|}, \frac{u_{ox}}{|\nabla u_o|}\right)$$

a unit vector in the s-direction. Therefore using the notation (2.8.6) we can write

$$(2.8.16) \qquad \varphi \sim \varphi_o + t\frac{\underset{\sim}{e}_u \cdot \nabla_o u}{|\nabla_o u|} + s^2(\underset{\sim}{e}_s, \underset{\sim}{\Phi}_o \underset{\sim}{e}_s)$$

Then since (2.8.14) and (2.8.15) define an orthogonal transformation $j_o = 1$, and inserting into (2.8.13) we have

$$R_{II} \sim e^{-z\varphi_o} \int_{-\infty}^{\infty} \alpha(s)ds \int_0^{\infty} \alpha(t)dt e^{-z\left(t\frac{\underset{\sim}{e}_u \cdot \nabla_o u}{|\nabla_o u|} + s^2(\underset{\sim}{e}_s, \underset{\sim}{\Phi}_o \underset{\sim}{e}_s)\right)}$$

and hence

$$(2.8.17) \qquad R_{II} \sim \frac{|\nabla_o u| e^{-z\varphi_o}}{z(\underset{\sim}{e}_u \cdot \underset{\sim}{\nabla}_o u_o)} \left(\frac{\pi}{(\underset{\sim}{e}_s, \underset{\sim}{\Phi}_o \underset{\sim}{e}_s)z}\right)^{\frac{1}{2}}$$

where the principal value square root is to be taken.

<u>Case 3</u>. Critical Point of Type III.

We consider

$$(2.8.18) \qquad R_{III} = \iint \alpha_3(x,y)\psi(x,y)e^{-z\varphi(x,y)}dxdy$$

where the critical point is a corner, type III.

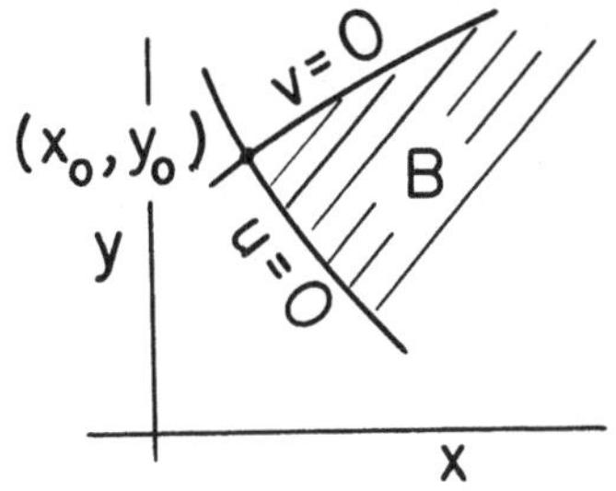

The corner point is described by the curves (see sketch above)

$$u(x,y) = 0$$
$$v(x,y) = 0$$

Also since (x_0, y_0) is a corner

$$\nabla u_0 \neq \nabla v_0$$

Next note that we may choose u and v so they increase on entering the domain B. (If any of these, say u, decrease we consider $-u$ instead of u). With these remarks in mind we use

$$u = u(x,y), \quad v = v(x,y)$$

as new coordinates in the neighborhood of the corner point. B transforms as is indicated in the sketch below,

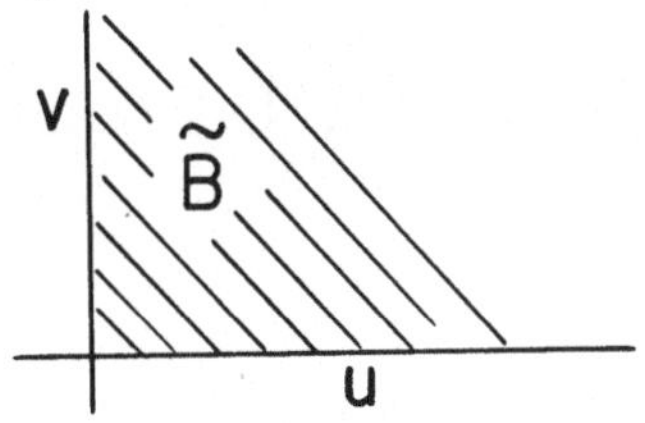

and (2.8.18) becomes

$$R_{III} = \iint \tilde{\alpha}(u,v)\tilde{\psi}(u,v)e^{-z\tilde{\varphi}(u,v)} \frac{\partial(x,y)}{\partial(u,v)} dudv$$

where for example $\tilde{\varphi}(u,v) = \varphi(x(u,v),y(u,v))$, $j = \frac{\partial(x,y)}{\partial(u,v)}$ is the functional determinant. In the neighborhood of the critical point $(u,v) = (0,0)$, we have

$$\tilde{\varphi} = \varphi_o + u\varphi_{ou} + v\varphi_{ov} + \dots$$

Again the formal expansion leads to the asymptotic expansion, so that

$$(2.8.19) \quad R_{III} \sim e^{-z\varphi_o} \int_0^\infty \alpha(u)du \int_0^\infty \alpha(v)dv(\psi_o+\dots)(j_o+\dots)e^{-z(u\tilde{\varphi}_{ou}+v\tilde{\varphi}_{ov})}(1+z\frac{u^2}{2}\tilde{\varphi}_{ouu}+\dots)$$

where once again we have used a product representation of the neutralizer. Once again we mention that the calculation of the terms of (2.8.19) follows that given in (2.3.10) and (2.4.13) and we do not carry the calculation further.

To calculate the lead term of (2.8.19) we note that

$$\underset{\sim}{e}_u = (u_{ox}, u_{oy})/|\nabla u_o|, \quad \underset{\sim}{e}_v = (v_{ox}, v_{oy})/|\nabla v_o|$$

are unit vectors in the u and v coordinate directions, respectively. From this

$$\varphi \sim \varphi_o + u\underset{\sim}{e}_u \cdot \underset{\sim}{\nabla}\varphi_o + v\underset{\sim}{e}_v \cdot \underset{\sim}{\nabla}\varphi_o + \dots$$

The ratio of areas is immediate and is

$$\frac{dudv}{dxdy} = \begin{vmatrix} u_{ox} & u_{oy} \\ v_{ox} & v_{oy} \end{vmatrix} / (|\nabla u_o||\nabla v_o|)$$

Hence

$$R_{III} \sim e^{-z\varphi_0} \int_0^\infty \alpha(u)du \int_0^\infty \alpha(v)dv\psi_0 \frac{|\nabla u_0||\nabla v_0|}{(u_{ox}v_{oy}-v_{ox}u_{oy})} \exp[-z(u\underset{\sim}{e}_u\cdot\nabla\varphi_0+v\underset{\sim}{e}_v\cdot\nabla\varphi_0)]$$

and

$$(2.8.20) \qquad R_{III} \sim \frac{\psi_0|\nabla u_0|^2|\nabla v_0|^2\exp(-z\varphi_0)}{z^2(u_{ox}v_{oy}-v_{ox}u_{oy})(u_{ox}\varphi_{ox}+u_{oy}\varphi_{oy})(v_{ox}\varphi_{ox}+v_{oy}\varphi_{oy})}$$

Exercise 46. Compute the leading term in the expansion of R when a type I and a type III point coincide.

2.9. Multidimensional Integrals: Part II. Many Parameters

We now consider the asymptotic approximation of integrals of the following type,

(2.9.1) $$A = \frac{1}{(2\pi)^N} \int_{-\infty}^{\infty} F(\underset{\sim}{x},\underset{\sim}{k}) e^{-\sigma(\underset{\sim}{k})t} \rho(\underset{\sim}{k}) d\underset{\sim}{k}$$

Both $\underset{\sim}{k}$ and $\underset{\sim}{x}$ represent real N-dimensional vectors, $d\underset{\sim}{k}$ represents the N-space volume element and F, σ, ρ are in general complex valued. The basic assumption on $F(\underset{\sim}{x},\underset{\sim}{k})$ is that it is uniformly bounded

(2.9.2) $$|F(\underset{\sim}{x},\underset{\sim}{k})| \leq L < \infty$$

in $\underset{\sim}{x}$ and $\underset{\sim}{k}$. We therefore include the important class of integrals

(2.9.3) $$B = \frac{1}{(2\pi)^N} \int_{-\infty}^{\infty} \rho(\underset{\sim}{k}) \exp(i\underset{\sim}{k}\cdot\underset{\sim}{x} - \sigma(\underset{\sim}{k})t) d\underset{\sim}{k}$$

which often arise in the solution of linear problems.

In regard to $\sigma(\underset{\sim}{k})$ we can without loss of generality take

(2.9.4) $$\sigma(\underset{\sim}{k} = 0) = 0$$

Further we will say that $\sigma(\underset{\sim}{k})$ is admissible if it satisfies the following five conditions:

(i) $\mathrm{Re}\,\sigma = \sigma_r \geq 0$

(ii) $\sigma_r = 0$ only if $k = |\underset{\sim}{k}| = 0$

(iii) $\sigma \in C$

(iv) in the neighborhood of the origin $\sigma = if(\underset{\sim}{k}) + g(\underset{\sim}{k}) + O(k^3)$

where f and g are real continuous, and homogeneous of degrees one and two respectively.

(v) $g = 0$, only if $\underset{\sim}{k} = 0$.

In brief, (i) states that $-\sigma_r$ has a global maximum (ii) that a "dissipative" mechanism is present. (The smaller wave number, $\underset{\sim}{k}$, contributions vanish least rapidly.) Condition (iv) is obtained if $\sigma \in C^3$ and is therefore somewhat weaker. (That the first order is pure imaginary and the second order pure real, generally follows from elementary physical considerations.)

Finally we take $\rho(\underset{\sim}{k})$ such that

$$|\rho|, \int_{-\infty}^{\infty} |\rho| d\underset{\sim}{k} < M < \infty \tag{2.9.5}$$

Theorem 291. For σ admissible, F satisfying (2.9.2) and ρ satisfying (2.9.5), (2.9.1) can be written in the limit $t \to \infty$, as

$$A = A_0 + O^*(t^{-(N+1)/2}) \tag{2.9.6}$$

where

$$A_0 = (2\pi)^{-N} \int_{-\infty}^{\infty} F(\underset{\sim}{x},\underset{\sim}{k})\rho(\underset{\sim}{k})\exp(-ift-gt)d\underset{\sim}{k} \tag{2.9.7}$$

The estimate O^* is such that $O^*(t^{-p}) = O(t^{-p+\delta})$ for any $\delta > 0$.

Corollary. If in addition to the hypothesis of Theorem 291 we have that

$$\rho = k^{\nu}(\rho_o + O(k)) \tag{2.9.8}$$

for small k, ($\nu \geq 0$ follows from (2.9.5)) then

$$A = A_0 + O^*(t^{-(N+1+\nu)/2}) \tag{2.9.9}$$

and

$$A = A^0\rho_0 + O^*(t^{-(N+1+\nu)/2}) \tag{2.9.10}$$

where

$$(2.9.11) \qquad A^{o} = (2\pi)^{-N} \int_{-\infty}^{\infty} F(\underset{\sim}{x},\underset{\sim}{k}) k^{\nu} \exp(-ift-gt) d\underset{\sim}{k}$$

In particular when $F = e^{i\underset{\sim}{k}\cdot\underset{\sim}{x}}$ we have

$$(2.9.12) \qquad B = B_{o} + O^{*}(t^{-(N+1)/2})$$

with

$$(2.9.13) \qquad B_{o} = (2\pi)^{-N} \int_{-\infty}^{\infty} \rho(\underset{\sim}{k}) \exp(i\underset{\sim}{k}\cdot\underset{\sim}{x}-ift-gt) d\underset{\sim}{k}$$

and when (2.9.8) is satisfied

$$(2.9.14) \qquad B = B_{o} + O^{*}(t^{-(N+1+\nu)/2})$$

and

$$(2.9.15) \qquad B = B^{o}\rho_{o} + O^{*}(t^{-(N+1+\nu)/2})$$

where

$$(2.9.16) \qquad B^{o} = (2\pi)^{-N} \int_{-\infty}^{+\infty} k^{\nu} \exp(i\underset{\sim}{k}\cdot\underset{\sim}{x}-ift-gt) d\underset{\sim}{k}$$

Before proving the various results we make some general comments. First we note that the result of Theorem 291 is in a sense only semi-constructive. Even the integral (2.9.16) cannot in general be carried out in terms of known functions -- in spite of the homogeneity requirements on f and g given in (iv). Secondly we note the somewhat passive role of $\underset{\sim}{x}$. The error estimates of (2.9.9), (2.9.10) etc., are independent of $\underset{\sim}{x}$. This too is an aspect of the semi-constructive nature of the calculation. The region of validity in $\underset{\sim}{x}$-space of the calculation is the restriction to those $\underset{\sim}{x}$ such that the integral terms in (2.9.9-10) and so forth, is large compared with the error estimate. The extent of the region follows from a study of the integral term -- often it requires the completion of the integration into known functions. In this same vein we note that it is conceivable that the integral terms will be less than or equal to the error estimate for all $\underset{\sim}{x}$. In such a case the calculation stands only as an estimate for the integral A, (2.9.1). Finally we for example distinguish between (2.9.9) and (2.9.10). In general the modulus of the error estimates in (2.9.9) and (2.9.10) are quite different. The former result is valid, usually, for significantly shorter times than the latter.

We will return to a number of these points in the examples given later.

In order to prove Theorem 291 and its corollary we first prove the following

Lemma. For $\sigma(\underset{\sim}{k})$ admissible there exist constants $g_0 > 0$ and $\varepsilon_1 > 0$ such that

$$\sigma_r - \hat{\varepsilon}^2 g_0 \geq 0$$

for all $|\underset{\sim}{k}| = k > \hat{\varepsilon}$ and any $\hat{\varepsilon} > 0$, such that $\hat{\varepsilon} \leq \varepsilon_1$.

Proof. Since g is homogeneous of degree two

$$g(\underset{\sim}{k}) = k^2 g(\underset{\sim}{e})$$

with

$$\underset{\sim}{e} = \frac{\underset{\sim}{k}}{k}$$

From the continuity of g and condition (v) we have

$$g_M \geq g(\underset{\sim}{e}) \geq g_m > 0$$

with g_M, g_m the maximum and minimum of $g(\underset{\sim}{e})$.

From condition (iv)

$$\sigma_r - \frac{k^2}{2} g_m \geq \frac{k^2 g_m}{2} + O(k^3)$$

Hence there exists an $\varepsilon_o > 0$ such that

(2.9.17) $$\sigma_r - \frac{k^2}{2} g_m > 0$$

for

$$k < \varepsilon_o$$

Let in fact ε_o be the maximum such value.

Next we assert that there exists an $\varepsilon_1 > 0$ such that for $0 < \varepsilon \leq \varepsilon_1$

$$\sigma_r(\underset{\sim}{k}) - \frac{1}{2} g_m \varepsilon^2 \geq 0, \quad k > \varepsilon$$

For if this were not true there would exist a null sequence $\{\varepsilon_i\}$, $\varepsilon_i > 0$, $\varepsilon_i \to 0$ and a corresponding set of points $\{\underset{\sim}{k}_i\}$ such that

(2.9.18) $$\sigma_r(\underset{\sim}{k}_i) < \frac{1}{2} g_m \varepsilon_i^2$$
$$k_i > \varepsilon_i \quad i = 1,\ldots$$

Also from (2.9.17) we can take $k_i > \varepsilon_o$ for otherwise we are led to

$$k_i^2 < \varepsilon_i^2$$

which is a contradiction. Denoting the limit point of the sequence $\{\underset{\sim}{k}_i\}$ by $\underset{\sim}{k}^*$, we have

$$\lim_{\underset{\sim}{k}_i \to \underset{\sim}{k}^*} \sigma_r(\underset{\sim}{k}_i) = 0$$

But from (iii) σ_r is continuous, and hence $\sigma_r(\underset{\sim}{k}^*) = 0$, which since $k^* \neq 0$ contradicts (ii).

Therefore taking

$$g_o = \frac{1}{2} g_m$$

we have proven the lemma.

Proof of Theorem 2.9.1.

From condition (iv) there exist constants $c > 0$ and $\varepsilon_2 > 0$ such that

$$|\sigma - if - g| < ck^3 \tag{2.9.19}$$

for

$$k < \varepsilon_2$$

Take $\varepsilon_3 = \min(\varepsilon_2, \varepsilon_1)$ where ε_1 is the same as that of the lemma.

For $\varepsilon < \varepsilon_3$ we decompose (2.9.1) as follows

$$\begin{aligned} A &= (2\pi)^{-N} \int_{k \geq \varepsilon} F(\underset{\sim}{x},\underset{\sim}{k})\rho(\underset{\sim}{k}) e^{-\sigma t} d\underset{\sim}{k} \\ &+ (2\pi)^{-N} \int_{k<\varepsilon} \rho(\underset{\sim}{k})F(\underset{\sim}{x},\underset{\sim}{k})[\exp(-\sigma t) - \exp(-ift-gt)]d\underset{\sim}{k} \\ &+ (2\pi)^{-N} \int_{k<\varepsilon} \rho(\underset{\sim}{k})F(\underset{\sim}{x},\underset{\sim}{k})\exp(-ift-gt)d\underset{\sim}{k} \\ &= A_1 + A_2 + A_3 \end{aligned} \tag{2.9.20}$$

From the lemma we obtain

$$(2.9.21)\qquad \begin{aligned} |A_1| &\le (2\pi)^{-N} L \int_{k\ge\varepsilon} e^{-\sigma_r t} |\rho|\, d\underset{\sim}{k} = (2\pi)^{-N} L e^{-g_o \varepsilon^2 t} \int_{k\ge\varepsilon} |\rho| \exp(g_o \varepsilon^2 t - \sigma_r t)\, d\underset{\sim}{k} \\ &\le (2\pi)^{-N} L M e^{-g_o \varepsilon^2 t} \end{aligned}$$

Considering A_2, we first write it as

$$A_2 = (2\pi)^{-N} \int_{k<\varepsilon} F(\underset{\sim}{x},\underset{\sim}{k})\rho(\underset{\sim}{k})\{\exp[(g+if-\sigma)t]-1\}\exp(-ift-gt)\, d\underset{\sim}{k}$$

and from this

$$|A_2| \le (2\pi)^{-N} LM \int_{k<\varepsilon} |\exp[(g+if-\sigma)t]-1|\, d\underset{\sim}{k}$$

$$|A_2| \le (2\pi)^{-N} LM \int_{k<\varepsilon} t|g+if-\sigma| \exp|(g+if-\sigma)t|\, d\underset{\sim}{k}$$

Therefore from (2.9.19)

$$|A_2| \le (2\pi)^{-N} LMct \int_{k<\varepsilon} k^3 e^{ck^3 t} d\underset{\sim}{k} = (2\pi)^{-N} LMc\Omega_N t \int_0^{\varepsilon} k^{N+2} e^{ck^3 t} dk$$

where

$$(2.9.22)\qquad \Omega_N = \frac{2\pi^{N/2}}{\Gamma(N/2)}$$

is the surface area of the unit sphere in N-dimensions. This then gives

$$(2.9.23)\qquad |A_2| \le \frac{(2\pi)^{-N} LMc\Omega_N t\, \varepsilon^{N+3}\, e^{c\varepsilon^3 t}}{N+3}$$

Finally we set

$$(2.9.24)\qquad \varepsilon = \frac{1}{t^{(1-\delta)/2}}$$

where $\delta > 0$ is as small as we please. We must therefore choose t sufficiently large so that $\varepsilon < \varepsilon_3$. With this choice of ε (2.9.21) becomes

$$|A_1| \le (2\pi)^{-N} LM \cdot e^{-g_0 t^{\delta}}$$

and (2.9.22)

$$(2.9.25) \quad |A_2| \le \frac{LM\Omega_N c}{(2\pi)^N (N+3)} \frac{t^{\frac{\delta}{2}(N+3)}}{t^{\frac{N+1}{2}}} \exp[ct^{-(1-3\delta)/2}] = O^*(t^{-(N+1)/2})$$

Hence since A_1 is exponentially decaying for $t \to \infty$, this proves our theorem.

The relations (2.9.9) and (2.9.10) of the corollary follow directly from the decomposition (2.9.20) and a similar set of estimates. This is easily seen and we leave the details as an exercise.

Complete Asymptotic Development

A many termed expansion can be also obtained with the above analysis. Let us suppose that instead of (iv) we have

$$\text{(iv)}' \qquad \sigma = \sum_{n=1}^{p} i^n \sigma_n + O(k^{p+1})$$

where the σ_n are homogeneous of degree n, real and continuous. We again write

$$A = A_1 + A_2 + A_3$$

where A_1 is as in (2.9.20) but we now take

$$(2.9.26) \qquad A_3 = (2\pi)^{-1} \int_{k<\varepsilon} F(\underset{\sim}{x},\underset{\sim}{k})\rho(\underset{\sim}{k}) \exp[t \sum_{n=1}^{p} i^n \sigma_n(\underset{\sim}{k})] d\underset{\sim}{k}$$

It then follows as before that

$$A_2 = O^*(t^{-(N+p-1)/2})$$

and

$$A = A_3 + O^*(t^{-(N+p-1)/2})$$

If $p > 2$ we cannot extend the limits of integration in (2.9.26) to infinity since the integral may not converge. To overcome this we first consider

$$e^{t(i\sigma_1 - \sigma_2)}[1 + t\sum_{n=3}^{p}\sigma_n i^n + \frac{1}{2!}(t\sum_{n=3}^{p}\sigma_n i^n)^2 + \dots$$

$$+ \frac{1}{(N+p-2)!}(t\sum_{n=3}^{p}\sigma_n i^n)^{N+p-2}]$$

and note that

$$|e^z - 1 - z - \frac{z^2}{2!} - \dots \frac{z^{p-1}}{(p-1)!}| \leq \frac{|z|^p e^{|z|}}{p!} \tag{2.9.27}$$

Hence

$$A_3 = \frac{1}{(2\pi)^N}\int_{k<\varepsilon} F(\underset{\sim}{x}\cdot\underset{\sim}{k})\rho(\underset{\sim}{k})e^{t(i\sigma_1-\sigma_2)}[1 + t\sum_{n=3}^{p}\sigma_n i^n + \dots$$

$$+ \frac{1}{(N+p-2)!}(t\sum_{n=3}^{p}\sigma_n i^n)^{N+p-2}]d\underset{\sim}{k} + O^*(t^{-(N+p-1)/2})$$

And finally

$$A = \frac{1}{(2\pi)^N}\int_{-\infty}^{\infty} F(\underset{\sim}{x}\cdot\underset{\sim}{k})g(\underset{\sim}{k})e^{t(i\sigma_1-\sigma_2)}[1 + \dots + \frac{1}{(N+p-2)!}(t\sum_{n=3}^{p}\sigma_n i^n)^{N+p-2}]d\underset{\sim}{k}$$

(2.9.28)

$$+ O^*(t^{-(N+p-1)/2})$$

(2.9.28) itself contains a number of higher order terms - but we do not discuss this further.

Examples

Before applying the above discussion it will be useful to make some preliminary remarks about Fourier integrals. We introduce the Fourier transform of $\rho(\underset{\sim}{k})$

$$\rho(\underset{\sim}{x}) = \frac{1}{(2\pi)^N} \int_{-\infty}^{\infty} e^{i\underset{\sim}{k}\cdot\underset{\sim}{x}} \rho(\underset{\sim}{k})d\underset{\sim}{k}$$

which exists by virtue of (2.9.5). (We often adopt the convention that the argument of a function indicates whether we are speaking of the function on its transform). Then it can be shown that if all relevant quantities exist,

$$\frac{1}{(2\pi)^N} \int_{-\infty}^{\infty} f(\underset{\sim}{k})g(\underset{\sim}{k})e^{i\underset{\sim}{k}\cdot\underset{\sim}{x}}d\underset{\sim}{k} = \int_{-\infty}^{\infty} f(\underset{\sim}{x}-\underset{\sim}{y})g(\underset{\sim}{y})d\underset{\sim}{y}$$

$$= \int_{-\infty}^{\infty} g(\underset{\sim}{x}-\underset{\sim}{y})f(\underset{\sim}{y})d\underset{\sim}{y} = f(\underset{\sim}{x})*g(\underset{\sim}{x})$$

(say by showing that both sides have the same Fourier transform). The product $f(\underset{\sim}{x})*g(\underset{\sim}{x}) = g(\underset{\sim}{x})*f(\underset{\sim}{x})$ is referred to as the convolution product.

From this it is tempting to write (2.9.3) as a convolution

$$\frac{1}{(2\pi)^N} \int_{-\infty}^{\infty} \exp[i\underset{\sim}{k}\cdot\underset{\sim}{x} - \sigma(\underset{\sim}{k})t]d\underset{\sim}{k}*\rho(x)$$

however the left member does not necessarily exist. To oversome this we define

(2.9.29)
$$\mathscr{J} = (2\pi)^{-N} \int_{k\leq 1} \exp[i\underset{\sim}{k}\cdot\underset{\sim}{x} - \sigma(\underset{\sim}{k})t]d\underset{\sim}{k}$$

then

$$B = \mathscr{J}*\rho + O(e^{-\bar{\sigma}t})$$

where $\bar{\sigma} = \min \operatorname{Re} \sigma(\frac{\underset{\sim}{k}}{k}) > 0$. The choice of $k = 1$ as the cutoff is arbitrary and plays no real role.

Next although the transform of $\exp(-\sigma(\underset{\sim}{k})t)$ may not exist the transform of $\exp(-if(\underset{\sim}{k})t - g(\underset{\sim}{k})t)$ will always exist for admissible σ. We therefore have

$$B = \mathcal{F}*\rho + O(e^{-\bar{\sigma}t}) = B^o(\underset{\sim}{x})*\rho(\underset{\sim}{x}) + O*(t^{-\frac{N+1}{2}})$$

and also

$$\mathcal{F} = B^o + O*(t^{-\frac{N+1}{2}})$$

Example. Suppose $F = e^{i\underset{\sim}{k}\cdot\underset{\sim}{x}}$ and that $\sigma(\underset{\sim}{k})$ has two derivatives. It then follows that

$$\sigma = i\underset{\sim}{\alpha}\cdot\underset{\sim}{k} + \underset{\sim}{k}\cdot\underset{\sim}{\beta}\cdot\underset{\sim}{k} + O(k^3) \tag{2.9.30}$$

where $\underset{\sim}{\alpha}$ is a real constant vector and $\underset{\sim}{\beta}$ is a real symmetric matrix of order N, which by (i) and (v) is positive definite. In this case

$$B^o = (2\pi)^{-N}\int_{-\infty}^{\infty}\exp(i\underset{\sim}{k}\cdot(\underset{\sim}{x}-\underset{\sim}{\alpha}t) - \underset{\sim}{k}\cdot\underset{\sim}{\beta}\cdot\underset{\sim}{k}t)d\underset{\sim}{k}\,. \tag{2.9.31}$$

One may then

Exercise 47. Show

$$B^o = \frac{\exp\left[-\dfrac{(\underset{\sim}{x}-\underset{\sim}{\alpha}t)\cdot\underset{\sim}{\beta}^{-1}\cdot(\underset{\sim}{x}-\underset{\sim}{\alpha}t)}{4t}\right]}{(4\pi t)^{N/2}[\det\underset{\sim}{\beta}]^{1/2}} \tag{2.9.32}$$

From (2.9.12) we can give a precise characterization to the region in $\underset{\sim}{x}$-space for which the asymptotic approximation is valid. For this purpose we consider (2.9.12) which is

$$B = B^o\rho_o + O*(t^{-(N+1)/2})$$

Hence the condition on $\underset{\sim}{x}$ is

$$(\underset{\sim}{x}-\underset{\sim}{\alpha}t)\cdot\underset{\sim}{\beta}^{-1}\cdot(\underset{\sim}{x}-\underset{\sim}{\alpha}t) = O(t \ln t)$$

Outside this region we have the estimate

$$B = O^*(t^{-(N+1)/2}).$$

Example. In a number of applications, due to isotropy, the admissible σ is a function only of the magnitude, k. Then although σ is not differentiable the condition (iv) implies the simple form

$$\sigma = i\alpha k + \beta k^2 + O(k^3)$$

Actually in order to illustrate some of our earlier statements we will assume instead

$$\sigma = i\alpha k + \beta k^2 + i\gamma k^3 + O(k^4) \tag{2.9.33}$$

We start by considering

$$\mathcal{J} = (2\pi)^{-N} \int_{k\leq 1} \exp(i\underset{\sim}{k}\cdot\underset{\sim}{x} - \sigma(\underset{\sim}{k})d\underset{\sim}{k}$$

Then since σ is only a function of k, one can

Exercise 48. Show

$$\mathcal{J} = \frac{1}{(2\pi)^{N/2} r^{(N/2)-1}} \int_0^1 J_{(N/2)-1}(kr)k^{N/2}e^{-\sigma(k)t}dk \tag{2.9.34}$$

and we consider (for $N \geq 2$)

$$\mathcal{J}(N) = \frac{1}{(2\pi)^{N/2}} \int_0^1 J_{(N/2)-1}(kr)k^{N/2}e^{-\sigma(k)t}dk$$

Define

$$A(N) = (2\pi)^{-N/2}\int_0^\infty J_{(N/2)-1}(kr)k^{N/2}e^{-i\alpha kt-\beta k^2 t}(1-i\gamma k^3 t)dk$$

It then follows from our earlier discussion namely (2.9.14) and (2.9.28) that

(2.9.35) $$\mathcal{J}(N) = A(N) + O^*(t^{-N/4 - 3/2})$$

We consider the case $N = 3$ first. Then it is well-known that

(2.9.36) $$J_{1/2} = (\frac{2}{\pi x})^{\frac{1}{2}}\sin x$$

and therefore

$$A(3) = \frac{1}{2\pi^2\sqrt{r}} \int_0^\infty k \sin kr\, e^{-i\alpha kt-\beta k^2 t}dk$$

$$- \frac{i\gamma t}{2\pi^2 \sqrt{r}} \int_0^\infty k^4 \sin kr\, e^{-i\alpha kt-\beta k^2 t}dt$$

Both integrals are standard and may be represented in terms of the parabolic cylinder function $D_\nu(z)$ [see e.g. Magnus, Oberhettinger, Soni, [14]].
In fact

$$A(3) = \frac{i}{8\pi^2 t\sqrt{r}}\{D_{-2}(\frac{i(r+\alpha t)}{(2\beta t)^{\frac{1}{2}}})\exp[- \frac{(r+\alpha t)^2}{8\beta t}] - D_{-2}(\frac{i(\alpha t-r)}{(2\beta t)^{\frac{1}{2}}}).$$

(2.9.37) $$\exp[- \frac{(r-\alpha t)^2}{8\beta t}]\} + \frac{3\gamma}{2\pi^2\beta t(2\beta tr)^{\frac{1}{2}}}\{D_{-5}(\frac{i(r+\alpha t)}{(2\beta t)^{\frac{1}{2}}})\exp[- \frac{(r+\alpha t)^2}{8\beta t}]$$

$$- D_{-5}(\frac{i(\alpha t-r)}{(2\beta t)^{\frac{1}{2}}})\exp[- \frac{(r-\alpha t)^2}{8\beta t}]\}$$

Although this does complete the calculation, a number of simplifying calculations may be carried out.

We use

(2.9.38) $$D_\nu(z) = e^{-z^2/4}(z^\nu + O(z^{\nu-2}))$$

First we note that if $r = o(t)$ then $r \underline{+} t >> 1$ for $t \to \infty$. Then using (2.9.38) we see that

$$A(3) = O(t^{-5/2})$$

which is smaller than the error estimate given in (2.9.35). Hence we may assume that $r \geq O(t)$. In any case $r+t >> 1$ as $t \to \infty$ and this then leads to

$$\mathscr{I}(N=3) = -\frac{i}{8\pi^2 tr} D_{-2}\left(\frac{i(\alpha t-r)}{(2\beta t)^{\frac{1}{2}}}\right) \exp\left[-\frac{(r-\alpha t)^2}{8\beta t}\right]$$

$$-\frac{3\gamma}{2\pi^2\beta tr(2\beta t)^{\frac{1}{2}}} D_{-5}\left(\frac{i(\alpha t-r)}{(2\beta t)^{\frac{1}{2}}}\right) \exp\left[-\frac{(r-\alpha t)^2}{8\beta t}\right] + O^*(t^{-11/4})$$

with the understanding that $r = O(t)$ and $\mathscr{I} = O^*(t^{-11/4})$ for $r = O(t)$.

The parabolic cylinder function may be replaced by the more familiar error function through

$$D_{-n-1}(z) = \left(\frac{\pi}{2}\right)^{\frac{1}{2}} \frac{(-1)^n}{n!} e^{-z^2/4} \frac{d^n}{dz^n}\left[e^{z^2/2}\operatorname{erfc}\left(\frac{z}{\sqrt{2}}\right)\right]$$

This leads to the useful expression

(2.9.40) $$\operatorname{Re}\mathscr{I} = \frac{e^{-\frac{(r-\alpha t)^2}{4\beta t}}}{16(\pi\beta t)^{3/2} r}\left\{(r-\alpha t) + \frac{\gamma}{2\beta}\left[\frac{(r-\alpha t)^4}{4\beta^2 t^2} - \frac{6(r-\alpha t)^2}{2\beta t} + 3\right]\right\} + O^*(t^{-11/4}).$$

Returning to the general case, arbitrary number of dimensions N, and taking only a two term expansion of σ, $\sigma = i\alpha k+\beta k^2+O(k^3)$, we can write

$$\mathcal{I} = B^0 + \frac{1}{r^{(N/2)-1}} O^*(t^{-(N/4)-1})$$

with

$$B^0 = \frac{1}{(2\pi)^{N/2} r^{(N/2)-1}} \int_0^\infty J_{(N/2)-1}(kr) k^{N/2} \exp(-i\alpha k-\beta k^2) dk \tag{2.9.41}$$

For all odd values of N equation (2.9.41) may be exactly evaluated. If N is even however no explicit expressions are available. Actually a second asymptotic analysis may be applied to (2.9.41), which yields an explicit form. We bypass this analysis (see L. Sirovich, Journal of Mathematical Physics, vol. 11, p. 1365, 1970) and quote the result

$$\mathcal{I}(N) = \frac{\Gamma(\frac{N+1}{4})\Gamma(\frac{N+3}{4})\exp[-\frac{i(N-1)\pi}{4} - \frac{(r-\alpha t)^2}{8\beta t}]}{2\Gamma(\frac{1}{2})\, p^{\frac{N}{2}+\frac{3}{2}}(2\pi^2\beta t)^{(N+1)/4}}$$

$$\cdot D_{-(N+1)/2}\left(\frac{i(\alpha t-r)}{(2\beta t)^{\frac{1}{2}}}\right) + O^*(t^{-\frac{3N}{4}}) \quad p \geq O(t) \tag{2.9.42}$$

$$\mathcal{I}(N) = O^*(t^{-\frac{3N}{4}}), \quad r = o(t).$$

<u>Application</u>

Consider the following initial value problem

$$\left(\frac{\partial^2}{\partial t^2} - \nabla^2 - \mu\nabla^2 \frac{\partial}{\partial t}\right)s = 0$$

$$s\big|_{t=0} = 2\delta(\underset{\sim}{x}), \quad \frac{\partial s}{\partial t}\Big|_{t=0} = \mu\nabla^2\delta(\underset{\sim}{x}) \tag{2.9.43}$$

$\delta(\underset{\sim}{x})$ is the Dirac delta "function". (The solution to this problem can be used to construct the solution to (2.9.43) with arbitrary data).

To solve we formally introduce the Fourier transforms

$$\hat{s}(\underset{\sim}{k},t) = \int_{-\infty}^{\infty} e^{-i\underset{\sim}{k}\cdot\underset{\sim}{x}} s(\underset{\sim}{x},t)d\underset{\sim}{x}$$

$$s(\underset{\sim}{x},t) = (2\pi)^{-N} \int_{-\infty}^{\infty} e^{i\underset{\sim}{k}\cdot\underset{\sim}{x}}\hat{s}(\underset{\sim}{k},t)d\underset{\sim}{k}$$

Fourier transforming the equation (2.9.43) and the data we obtain

$$\frac{\partial^2}{\partial t^2}\hat{s} + k^2\hat{s} + \mu k^2 \frac{\partial\hat{s}}{\partial t} = 0$$

$$\hat{s}\Big|_{t=0} = 2, \quad \frac{\partial\hat{s}}{\partial t}\Big|_{t=0} = -\mu k^2$$

$$\hat{s} = e^{\sigma_+(k)t} + e^{\sigma_-(k)t}$$

where

$$\sigma_{\pm} = \frac{1}{2}(-\mu k^2 \pm (\mu^2 k^4 - 4k^2)^{\frac{1}{2}})$$

The solution to the problem (2.9.43) is formally given by

$$s(\underset{\sim}{x},t) = s_+(\underset{\sim}{x},t) + s_-(\underset{\sim}{x},t)$$

with

$$s_{\pm}(\underset{\sim}{x},t) = (2\pi)^{-N} \int_{-\infty}^{\infty} \exp(i\underset{\sim}{k}\cdot\underset{\sim}{x} + \sigma_{\pm}(k)t)d\underset{\sim}{k}$$

As is easily seen both $\sigma_{\pm}$ effectively satisfy the admissibility requirements and in the neighborhood of the origin

(2.9.44) $$\sigma_{\pm} = \pm ik - \frac{\mu k^2}{2} \mp \frac{i\mu^2 k^3}{8} + O(k^4)$$

Since the solution $s(x,t)$ is real we need only compute $\text{Re}\, s_+$ and

Re s_-.

We now restrict ourselves to the case $N = 3$. The integrals s_+ and s_- under (2.9.44) are explicitly given by (2.9.40) and we have

$$s = \frac{\exp[-\frac{(r-t)^2}{2\mu t}]}{2(2\pi\mu t)^{3/2} r}\{(r-t) + \frac{\mu}{8}[\frac{(r-t)^4}{\mu^2 t^2} - \frac{6(r-t)}{\mu t} + 3]\}$$

$$+ O^*(t^{-11/4})$$

One may verify that in the limit $\mu \to 0$, this becomes

$$\frac{1}{2\pi}\frac{\delta'(r-t)}{r}$$

which is the solution to (2.9.42) with $\mu = 0$.

2.10. Asymptotic Evaluation of Integrals Involving Non-Uniformities

Consider the integral

(2.10.1) $$H(t,y) = \int_y^b g(x)\exp[t\ f(x)]dx, \quad t \to \infty$$

in which g and f are real and sufficiently smooth. The function $f(x)$ is assumed to have a single global maximum at $x = 0 < b$; where

$$f(0) = f^o, \quad f'^o = 0, \quad f''^o \neq 0$$

Then from Laplace's formula (2.3.2) we obtain;

$$H \sim g^o\left(\frac{2\pi}{-f''_o t}\right)^{\frac{1}{2}}\exp[t\ f^o], \quad y < 0$$

$$H \sim g^o\left(\frac{\pi}{-2f''_o t}\right)^{\frac{1}{2}}\exp[t\ f^o], \quad y = 0$$

and by direct parts integration,

$$H \sim \frac{g(y)}{f'(y)t} \exp[f(y)t], \quad y > 0$$

assuming that $f(y)$ is monotonic decreasing for $0 \leq y \leq b$.

This simple discussion illustrates the non-uniform behavior of an integral having an endpoint in the neighborhood of the global (sharp) maximum of the integrand.

A number of such non-uniformity problems occur in practice. In this section we shall give two illustrations of a general technique for dealing with problems of this type.

Integrals Containing a Global Maximum Near an Endpoint.

We consider

$$I(t;y) = \int_a^b g(x) \exp[tf(x;y)]dx \tag{2.10.2}$$

for $t \to \infty$. The parameter y represents the single maximum (global) of f;

$$f'(y;y) = 0, \quad f''(y;y) < 0$$

which may lie outside the interval of integration. [If we set $p = \frac{x-y}{b-y}$ in (2.10.1) we obtain an integral of the type (2.10.2)]. To be definite we shall regard y as being in the neighborhood of a. Also in the interest of generality we permit $g(x)$ to be singular at $x = a$, and we write

$$g = x^{-r}h(x), \quad r < 1$$

and h is sufficiently smooth.

The method of dealing with problems of this types lies in first finding a simple comparison integral which incorporates all the basic features of the problem. Thus in relation to (2.10.2) we can consider

(2.10.3) $$S(t;y) = \int_0^\infty e^{-t(\frac{x^2}{2}+yx)} \frac{dx}{x^r}$$

with $r < 1$. The function

$$f(x;y) = -\frac{x^2}{2} - xy$$

has the single maximum point at $x = y$ and $f''(y;y) = -1$. It is clear that $S(t;y)$ contains all the complexities of the above stated problem.

Exercise 49. Find the asymptotic expansions of $S(t,y)$ for $|y\sqrt{t}|$ large and small.

Next setting $\sqrt{t}\,x = p$ in (2.10.3) we obtain

(2.10.4)
$$\begin{aligned} S(t;y) &= t^{\frac{r-1}{2}} \int_0^\infty p^{-r} \exp[-\frac{p^2}{2} - y\sqrt{t}p]dp \\ &= t^{\frac{r-1}{2}} \Gamma(r-1)\exp[\frac{ty^2}{4}]D_{r-1}(y\sqrt{t}) \end{aligned}$$

where $D_\nu(z)$ is the parabolic cylinder function (see [14], chapter VII). Our objective is reduce (2.10.2) under the limit $t \to \infty$, to integrals of the form (2.10.3) and hence to parabolic cylinder functions - which are tabulated.

We may without loss of generality take $a = 0$ in (2.10.2), and also for convenience we will write t^2 instead of t. Therefore we consider

(2.10.5) $$I(t;y) = \int_0^b x^{-r}h(x)\exp[t^2 f(x,y)]dx$$

First we transform the exponential by

(2.10.6) $$f(x;y) - f(0,y) = -(\frac{p^2}{2} + \gamma p)$$

which determines

$$p = p(x) \quad \text{or} \quad x = x(p)$$

in which we take

$$x(0) = 0$$

Differentiating (2.10.6), we get

$$\frac{dx}{dp} = -\frac{p+\gamma}{f'(x;y)} \tag{2.10.7}$$

Hence we take

$$p(y) = -\gamma$$

in order to have (2.10.6) a one-to-one transformation. Next in order to evaluate (2.10.7) at $x = y$ $(p = -\gamma)$ we apply l'Hospital's rule

$$\left(\frac{dx}{dp}\right)_{p=-\gamma} = \frac{-1}{f''(y;y)\left(\frac{dx}{dp}\right)_{p=-\gamma}} = \pm(-1/f''(y;y))^{\frac{1}{2}}$$

and without loss of generality we may take the positive branch, so that

$$\left(\frac{dx}{dp}\right)_{p=-\gamma} = (-1/f''(y;y))^{\frac{1}{2}} \tag{2.10.8}$$

To determine γ we evaluate (2.10.6) at $x = y$ $(p = -\gamma)$, which gives

$$\gamma = \pm(2[f(y;y) - f(0;y)])^{\frac{1}{2}}$$

The choice of branch is best determined by examining the derivative of (2.10.6) at $x = 0$ $(p = 0)$,

(2.10.9) $$f'(0;y)\left(\frac{dx}{dp}\right)_{p=0} = -\gamma$$

Since we have fixed $\frac{dx}{dp} > 0$ and since $f'(0;y) > 0$ for $y > 0$; $f'(0;y) < 0$, $y < 0$ we conclude that $\operatorname{sgn} \gamma = -\operatorname{sgn} y$ and hence

(2.10.10) $$\gamma = -\operatorname{sgn} y\, [2(f(y;y) - f(0;y))]^{1/2}$$

Returning to (2.10.5), I now has the form

(2.10.11) $$I(t;y) = \int_0^{p(b)} [x(p)]^{-r} \frac{h(x(p))(p+\gamma)}{-f'(x(p);y)} \exp[-t^2(\frac{p^2}{2} + \gamma p)]dp e^{t^2 f(0,y)}$$

On introducing a Heaviside function at the upper limit we can extend the upper limit to ∞ and we assume this in the following. It is next clear that there is no simple expansion of the coefficient functions of the exponential in (2.10.11). For if $p = -\gamma$ is an interior point an expansion about $p = -\gamma$ is in order otherwise an expansion about $p = 0$ is called for. To take care of this we write,

(2.10.12) $$\frac{h(x)}{x^r}\frac{dx}{dp} = -\frac{1}{[x(p)]^r}\frac{h(x(p))}{f'(x(p))}(p+\gamma) = \frac{\alpha_0+\beta_0 p+p(p+\gamma)H_0(p)}{p^r}$$

Therefore

$$I = I_0 + I_1$$

(2.10.13) $$= e^{t^2 f(0;y)}\int_0^\infty \frac{(\alpha_0+\beta_0 p)}{p^r} \exp[-t^2(\frac{p^2}{2} + \gamma p)]dp$$

$$+ e^{t^2 f(0;y)}\int_0^\infty \frac{p(p+\gamma)H_0(p)}{p^r} \exp[-t^2(\frac{p^2}{2} + \gamma p)]dp$$

Next we define

(2.10.14) $$W_r(z) = \int_0^\infty p^{-r} \exp[-(\frac{p^2}{2} + zp)]dp$$

which is easily related to the parabolic cylinder function (2.10.4). Noting that

$$W'_r(z) = W_{r-1}$$

we can write

$$I_o = \{\frac{\alpha_o}{t^{1-r}} W_r(\gamma t) + \frac{\beta_o}{t^{2-r}} W'_r(\gamma t)\}\exp[t^2 f(0;y)]$$

Also

$$I_1 = -\frac{1}{t^2}\int_0^\infty \frac{H_o(p)}{p^{r-1}} \frac{d}{dp} \exp[-t^2(\frac{p^2}{2} + \gamma p)]dp \exp[t^2 f(0,y)]$$

$$= \frac{1}{t^2}\int_0^\infty \frac{(1-r)H_o(p)+H'_o(p)p}{p^r}\exp[-t^2(\frac{p^2}{2} + \gamma p)]dp \exp[t^2 f(0;y)]$$

This integral is in the same form as (2.10.11) and we repeat the process, i.e., we write

$$(1-r)H_o(p) + H'_o(p)p = \alpha_1 + \beta_1 p + p(p+\gamma)H_1(p)$$

From this it follows that

$$I_1 = \frac{1}{t^2}[\frac{\alpha_1 W_r(t\gamma)}{t^{1-r}} + \frac{\beta_1 W'_r(t\gamma)}{t^{2-r}}]\exp[t^2 f(0,y)]$$
$$+ \frac{I_2}{t^2} .$$

Therefore by repeated use of this procedure we find

(2.10.15) $$I(t;y) \sim \exp[t^2 f(0;y)]\{\frac{W_r(t\gamma)}{t^{1-r}} \sum_{i=0} \frac{\alpha_i}{t^{2i}} + \frac{W'_r(t\gamma)}{t^{2-r}} \sum_{i=0} \frac{\beta_i}{t^{2i}}\}$$

It remains for us to find the α_i and β_i. Setting $p = 0$ in (2.10.12) we have

$$\alpha_0 = \left(\frac{p}{x(p)}\right)^r_{p=0} h(0)\left(\frac{dx}{dp}\right)_{p=0}$$

$$= h(0)\left(\frac{dx}{dp}\right)^{1-r}_{p=0}$$

Hence from (2.10.9) and (2.10.10)

(2.10.16)
$$\alpha_0 = h(0)\left[\frac{\operatorname{sgn} y[2(f(y;y) - f(0;y)]^{1/2}}{f'(0,y)}\right]^{1-r}$$

Next setting $p = -\gamma$ $(x = y)$ in (2.10.12) we find

(2.10.17)
$$\beta_0 = \frac{h(y)}{(-\gamma)^{1-r} y^r (-f''(y;y))^{\frac{1}{2}}} + \frac{\alpha_0}{\gamma}$$

This in turn determines $H_0(p)$, and so on.

Exercise 50. Consider

$$\int_\beta^\infty \frac{\exp[-\eta(\tau + \frac{1}{\tau})]}{\tau(\tau-\beta)^{\frac{1}{2}}}$$

for $\eta \to \infty$ and all $\beta > 0$.

The above treatment extends to the methods of stationary phase and steepest descents. An exposition of these and other problems, as well as the discussion showing that (2.10.15) is the AD of I is to be found in [N. Bleistein, Communications on Pure and Applied Mathematics XIX, 4(1966)]. Earlier references are also given there.

Neighboring Saddle Points

In exercise 42 we encountered the Bessel function representation

(2.10.18)
$$J_\nu(\nu a) = \int_{\infty-i\pi}^{\infty+i\pi} \exp[\nu a \sin hz - \nu z]dz$$

The solution to that problem shows that if $\nu \to \infty$ and $0 < a < 1$ the main contribution comes from two saddle points. While when $a = 1$ the main contribution comes from a saddle of order 3 - a "monkey" saddle. To obtain a uniform approximation when $\nu \to \infty$ we must therefore consider two saddles of index 2 coalescing into a saddle of order three. The general investigation of this problem is to be found in [Chester, Friedman, Ursell, Proc. Comb. Phil Soc. 53, 599(1957); Friedman, J. Soc. Ind. Appl. Math. 3, 280(1955); Ursell, Proc. Comb. Phil. Soc. 61, 113 (1965)] as well as the above cited work of Bleistein.

As already mentioned the trick in dealing with non-uniformities lies in finding a simple comparison form which incorporates all the features of the problem at hand. It is clear that in the present class the proper comparison integral is

$$\int_\Gamma \exp[t(\frac{z^3}{3} - \alpha z)]$$

with a path Γ insuring convergence. The exponential has two simple saddle points at $\pm\sqrt{\alpha}$ which coalesce into a monkey saddle when $\alpha = 0$. On the other hand, this integral, under the transformation $t^{1/3}z = \eta$ is related to the Airy integrals discussed in section 2.7 . (We now assume the path Γ to be one of those used in section 2.7 .) The strategy is therefore to reduce (2.10.18) in the limit $t \to \infty$ to integrals of the Airy type.

Instead of dealing with (2.10.18) we proceed more generally. Consider

$$Y(t;\alpha) = \frac{1}{2\pi i} \int_\gamma g(z)\exp[tf(z;\alpha)]dz \tag{2.10.19}$$

with $g(z)$ and $f(z)$ analytic in the domain of interest. Furthermore

$$f'(z_1(\alpha);\alpha) = f'(z_2(\alpha);\alpha) = 0$$

and

$$f''(z_{1,2}(\alpha);\alpha) \neq 0, \quad \alpha \neq 0$$

When $\alpha = 0$

$$z_1(0) = z_2(0) = z_o$$

and we assume

$$f'''(z_o;0) \neq 0$$

<u>Exercise 51</u>. Prove $f''(z_o;0) = 0$.

We will assume that $f(z,\alpha)$ is analytic in a sufficiently large domain; for all values of the parameter α of interest. Further we assume that γ is a suitable path of integration, stretching across the saddle points and ending in the valleys of $f(z;\alpha)$. We only give a formal sketch of the asymptotic development of (2.10.19) and therefore avoid specifying detailed conditions on f, g and γ.

Following the plan described above we transform the integral (2.10.19) by setting

$$f(z;\alpha) = -\frac{u^3}{3} + \beta^2(\alpha)u + \nu(\alpha) \tag{2.10.20}$$

where the "constants" $\beta(\alpha)$, $\nu(\alpha)$ are still to be specified. Differentiating (2.10.20) we obtain

$$\frac{df}{dz}\frac{dz}{du} = -u^2 + \beta^2$$

or

$$\frac{dz}{du} = -\frac{(u^2-\beta^2)}{\frac{df}{dz}} \tag{2.10.21}$$

Therefore in order to have a regular, one-to-one, mapping we take

$$u(z_1;\alpha) = \beta(\alpha)$$

$$u(z_2;\alpha) = -\beta(\alpha)$$

Substituting these into (2.10.20) gives

$$f(z_{1,2};\alpha) = \pm \frac{2\beta^3}{3} + \nu$$

and therefore

$$\nu = \frac{f(z_1,\alpha) + f(z_2,\alpha)}{2}$$

(2.10.22)

$$\beta^3 = \frac{3}{4}\{f(z_1;\alpha) - f(z_2,\alpha)\}$$

Returning to (2.10.21) we can evaluate $\frac{dz}{du}$ at the saddle points by using l'Hospital's rule,

$$\left(\frac{dz}{du}\right)^2_{\pm\beta} = \frac{\mp 2\beta(\alpha)}{\frac{d^2f}{dz^2}(z_{1,2};\alpha)} , \quad \alpha \neq 0$$

and

$$\left(\frac{dz}{du}\right)^3_{u=0} = \frac{-2}{\frac{d^3f}{dz^3}(z_0;0)} , \quad \alpha = 0$$

Although we now have a number of necessary conditions, it has not been demonstrated that there exists a regular branch, $u(z)$, of (2.10.20) having all the desired properties. A proof of this under certain circumstances is to be found in the above mentioned article by Chester, Friedman and Ursell (a simpler version is in the above mentioned article by Friedman).

On substituting (2.10.20) into (2.10.19) we get

(2.10.23) $$Y(t;\alpha) = \frac{\exp(\nu t)}{2\pi i} \int_{\hat{\gamma}} h(u)\exp[-(\frac{u^3}{3} - \beta^2 u)t]du$$

where

$$h(u) = g(z(u))\frac{dz}{du}$$

and $\hat{\gamma}$ is the image path, which again runs from valley to valley across the saddle points.

Next we write

$$h = p_o + u\, q_o + (u^2-\beta^2)r(u) \tag{2.10.24}$$

and substituting this into (2.10.23) yields

$$Y = p_o Y_o + q_o Y_1 + Y_r$$

where

$$Y_o = \frac{1}{2\pi i}\int_{\hat{\gamma}} \exp[-(\frac{u^3}{3} - \beta^2 u)t]du \exp(\nu t)$$

and

$$Y_1 = \frac{1}{2\pi i}\int_{\hat{\gamma}} u\exp[-(\frac{u^3}{3} - \beta^2 u)]du \exp(\nu t)$$

Since the endpoints of $\hat{\gamma}$ lie in valleys we have

$$Y_o \sim \frac{\exp(\nu t)}{2\pi i}\int_{\Gamma} \exp[-(\frac{u^3}{3} - \beta^2 u)t]du = \frac{\exp(\nu t)}{t^{1/3}}A_i(\beta^2 t^{2/3})$$

where Γ denotes the appropriate Airy integral path and A_i the corresponding Airy function. Similarly

$$Y_1(t;\alpha) \sim \frac{\exp(\nu t)}{t^{2/3} 2\pi i} \int_\Gamma \eta \exp[-(\frac{\eta^3}{3} - \beta^2 t^{2/3}\eta]d\eta = \frac{\exp(\nu t)}{t^{2/3}} A_i'(\beta^2 t^{2/3})$$

where $A_i'(z) = \dfrac{dA_i}{dz}$.

Finally considering the remaining term

$$Y_r = \frac{\exp(\nu t)}{2\pi i} \int_{\hat{\gamma}} (u^2-\beta^2) r(u) \exp[-(\frac{u^3}{3} - \beta^2 u)t]du$$

$$= -\frac{\exp(\nu t)}{2\pi i} \cdot \frac{1}{t} \int_{\hat{\gamma}} r(u) \frac{d}{du}\{\exp[-(\frac{u^3}{3} - \beta^2 u)t]\}du$$

and since the endpoints of $\hat{\gamma}$ lie in valleys

$$Y_r(t;\alpha) \sim \frac{\exp(\nu t)}{2\pi i t} \int_{\hat{\gamma}} r'(u) \exp[-(\frac{u^3}{3} - \beta^2 u)t]du$$

which is t^{-1} times an integral of the form (2.10.23). The same procedure may again be applied to $r'(u)$. We do not pursue this further, but merely note that by these methods it is clear that we obtain

$$Y \sim \exp(\nu t)\{\frac{A_i(t^{2/3}\beta^2)}{t^{1/3}} \sum_{n=0} \frac{p_n}{t^n} + \frac{A_i'(t^{2/3}\beta^2)}{t^{2/3}} \sum_{n=0} \frac{q_n}{t^n}\}$$

The constants p_n and q_n are to be determined successively. For example from (2.10.24) we find

$$h(\pm\beta) = p_o \pm \beta q_o$$

so that

$$p_o = \frac{h(\beta) + h(-\beta)}{2}$$

$$q_o = \frac{h(\beta) - h(-\beta)}{2\beta}$$

Exercise 52. Find the first two terms in the uniform expansion of $J_\nu(\nu a)$, (2.10.18), as $\nu \to \infty$.

2.11. Miscellaneous.

Laplace Transforms in the Neighborhood of the Origin

From Watson's lemma we have that the Laplace transform

$$(2.11.1) \qquad \mathcal{L}(f) = \int_0^\infty e^{-zx} f(x)\,dx, \quad z \in S^\delta$$

depends on the behavior of $f(x)$ in the neighborhood of the origin. It is therefore tempting to suppose that $\mathcal{L}(f)$ for $z \sim 0$ depends on the behavior of $f(x)$ for $x \sim \infty$. This however is not entirely true. Suppose for example that

$$f = \beta x^\alpha + g(x), \quad 0 > \text{Re}\,\alpha > -1$$

with $g(x)$ absolutely integrable. Then it is clear that

$$\mathcal{L}(f) = \frac{\beta\alpha!}{z^{\alpha+1}} + O(1)$$

so that the singular behavior of $\mathcal{L}(f)$ depends on $f(x)$ for $x \sim \infty$. But it is also clear that any higher orders depend on $f(x)$ in the entire range of integration. As another illustration consider

$$(2.11.2) \qquad \mathcal{F}(z) = \int_0^\infty \frac{e^{-zt}}{1+t}\,dt \left(= \int_z^\infty e^{z-\zeta}\frac{d\zeta}{\zeta}\right)$$

Parts integrating we obtain,

$$\mathcal{F}(z) = z\int_0^\infty e^{-zt}\ell n(1+t)\,dt$$

$$= \alpha(z) + \beta(z)$$

where

$$\alpha(z) = \int_0^{-1} e^{-zt} \ell n(1+t) dt.$$

This has an integrable singularity and clearly is analytic at the origin (in fact it is entire). On the other hand if in the remaining integral

$$\beta = z\int_{-1}^{\infty} e^{-zt} \ell n(t+1) dt$$

we set $t = -1 + \xi/z$, we obtain

$$\beta(z) = e^z \int_0^{\infty} e^{-\xi}(\ell n\ \xi - \ell n\ z) d\xi$$

$$= e^z(\gamma - \ell n\ z)$$

with

$$\gamma = \int_0^{\infty} e^{-x} \ell n\ x\ dx$$

Euler's constant. The entire interval contributes in addition a logarithmic branch.

Exercise 53 . Find the expansion of $\mathcal{L}(f)$, $z \to 0$, $z \in S^{\delta}$ when

$$f(x) = \begin{cases} 0 & ; 0 \leq x < 1 \\ x^{-n}, \ n > 0; & x \geq 1. \end{cases}$$

From the above discussion we can infer that if

(2.11.3)
$$f = \sum_{i=1}^{n} \beta_i t^{\alpha_i} + r(t), \quad r = o(t^{\alpha_n})$$

with

$$\operatorname{Re} \alpha_1 \geq \operatorname{Re} \alpha_2 \geq \dots \geq \operatorname{Re} \alpha_n > -1$$

then

$$(2.11.4) \qquad \mathcal{L}(f) = \sum_{i=1}^{n} \frac{\beta_i \alpha_i!}{z^{\alpha_i+1}} + o(z^{-(\alpha_i+1)})$$

This will follow from the result which is obtained below.

Before coming to this we remark that more generally, if $r(t)$ in (2.11.3) is integrable then for $z \to 0$, $z \in S^{\delta}$,

$$(2.11.5) \qquad \mathcal{L}(f) = \sum_{i=1}^{n} \frac{\beta_i \alpha_i!}{z^{\alpha_i+1}} + h(z)\ell n\, z + g(z)$$

where $g(z)$ and $h(z)$ are analytic at the origin. The details of $g(z)$ and $h(z)$ as indicated above depend on $f(t)$ in the entire interval of integration. (2.11.2) is an illustration of (2.11.5). [This result under various assumptions is derived in the notes by Friedrichs [1], in Beyer and Heller, Jour. Math. Physics, 5, 1004 (1967), and in Handelsman and Lew, SIAM J. Math. Anal 1 (1970)]. The last reference contains a very general approach to this problem and its results seem to be most comprehensive.]

We now consider (2.11.1) for $z \to 0$, $z \in S_{\delta}$ and obtain the leading term in the AD of $\mathcal{L}(f)$ under somewhat more general circumstances. In particular we consider $f(x)$ such that

$$(2.11.6) \qquad \frac{f'}{f} \sim \frac{\mu}{x}, \quad \text{Re}\,\mu > -1$$

It then follows from our earlier analysis that

$$(2.11.7) \qquad f(x) = x^{\mu+o(1)}$$

(see pp. 45-47 for this derivation. Note on these pages it was assumed that f is positive and μ real, but on returning to Theorem 126, part A it is seen that these requirements are not necessary in the case under study.)

In (2.11.7), $o(1)$ represents a quantity which vanishes as $x \to \infty$. In particular

$$f(\tfrac{1}{t}) = (\tfrac{1}{t})^{\mu+o(1)}$$

$$f(\tfrac{x}{t}) = (\tfrac{x}{t})^{\mu+o(1)}$$

and therefore for x fixed and $t \to 0$,

$$f(\tfrac{x}{t}) \sim x^{\mu} f(\tfrac{1}{t})$$

or

$$f(\tfrac{x}{t}) = x^{\mu}[f(\tfrac{1}{t}) + o(f(\tfrac{1}{t}))] \tag{2.11.8}$$

Returning to (2.11.1) we write

$$z = te^{i\theta}, \quad |z| = t$$

and

$$\mathcal{L}(f) = \int_0^{\mathcal{E}(t)} e^{-zx} f(x)dx + \int_{\mathcal{E}(t)}^{\infty} e^{-zx} f(x)dx \tag{2.11.9}$$

where $\mathcal{E}(t)$ is such that for $t \to 0$

$$\begin{aligned} &\mathcal{E}(t) \to \infty \\ &t\mathcal{E}(t) \to 0 \end{aligned} \tag{2.11.10}$$

Setting $xt = s$ in the second integral of (2.11.9),

$$\int_{\mathcal{E}(t)}^{\infty} e^{-zx} f(x)dx = \frac{1}{t}\int_{t\mathcal{E}(t)}^{\infty} f(\tfrac{x}{t}) \exp[-e^{i\theta}x]dx$$

Next from (2.11.7) we see that for x large the neighborhood of ∞ contributes negligibly to the integral. Therefore we may replace ∞ at the upper limit by a suitably large constant. This done we can substitute (2.11.8) and return the upper limit to ∞, so finally

$$\int_{\mathcal{E}(t)}^{\infty} e^{-zx} f(x)dx \sim \frac{f(\frac{1}{t})}{t} \int_{t\mathcal{E}(t)}^{\infty} x^{\mu} \exp[-e^{i\theta}x]dx$$

Extending the lower limit contributes a negligible order and hence

$$(2.11.11) \qquad \int_{\mathcal{E}(t)}^{\infty} e^{-zx} f(x)dx \sim \frac{\mu! f(\frac{1}{|z|})}{z} e^{-i\mu\theta}$$

Next consider the first part of (2.11.9) we see by elementary estimates,

$$\int_{0}^{\mathcal{E}(t)} e^{-zx} f(x)dx = O(\mathcal{E}f(\mathcal{E}))$$

which from (2.11.7) is small compared to (2.11.11). [E.g. take $\mathcal{E} = O(\sqrt{t})$.] We have therefore proven the following:

For $f(x)$ such that

$$\frac{f'(x)}{f(x)} \sim \frac{\mu}{x}, \quad \operatorname{Re} \mu > -1$$

then

$$(2.11.12) \qquad \mathcal{L}(f) \sim \frac{\mu! \exp[-i\mu \arg z]}{z} f(\frac{1}{|z|})$$

for $z \in S^{\delta}$ and $z \to \infty$. (This demonstrates (2.11.4) under (2.11.3))

The case $\operatorname{Re} \mu = -1$ is not generally difficult but must be treated individually in each case. A case in point is contained in the following:

<u>Exercise 54.</u> For

$$f \sim \frac{c(\ln x)^{\beta}}{x}$$

$|c| \neq 0$ demonstrates

$$\mathcal{L}(f) \sim \begin{cases} \frac{c}{\beta+1}(-\ln t)^{\beta+1}, & \beta \neq -1 \\ c \ln|\ln t| & , \beta = -1 \end{cases}$$

$z = t$, real and $t \to 0$. [Hint: for $\beta \geq 0$ take $\varepsilon = O(\ln t)$ and for $\beta < 0$ take $\varepsilon = O(\sqrt{t})$]

Fourier Transforms at the Origin

We consider

$$F(f) = \int_0^\infty e^{itx} f(x)dx \tag{2.11.13}$$

under the assumption that

$$\frac{f'}{f} \sim \frac{\mu}{x}, \ -1 < \operatorname{Re} \mu < 0 \tag{2.11.14}$$

Then proceeding as above we write

$$F(f) = \int_0^{\varepsilon(t)} e^{itx} f(x)dx + \int_{\varepsilon(t)}^\infty e^{itx} f(x)dx \tag{2.11.15}$$

with $\varepsilon(t)$ chosen as in (2.11.10). Then setting $xt = s$ in the second integral

$$\int_{\varepsilon(t)}^\infty e^{itx} f(x)dx = \frac{1}{t}\int_{t\varepsilon(t)}^\infty e^{is} f(\tfrac{s}{t})ds$$

From (2.11.7) and (2.11.14) we have that this integral converges no matter how small t. The argument is as before and we can substitute (2.11.8) and hence

$$\frac{1}{t}\int_{t\varepsilon(t)}^\infty e^{is} f(\tfrac{s}{t})ds \sim \frac{1}{t} f(\tfrac{1}{t})\int_{t\varepsilon(t)}^\infty e^{is} s^\mu ds$$

The lower limit can be extended to the origin and the first integral of (2.11.15) is also negligible. Hence

$$F(f) \sim \frac{i\mu!}{t} f(\tfrac{1}{t}) e^{i\mu\pi/2} \tag{2.11.16}$$

which is (2.11.12) under the conditions of (2.11.13, 14).

Exercise 55. The Hilbert transform of $f(t)$ is defined as

$$\mathscr{H}(f) = \int_0^\infty \frac{f(t)}{t-z}\, dt$$

Under suitable conditions on $f(t)$ find the AD of $\mathscr{H}(f)$ for $|z|$ large and small. [Hint: Note

$$\frac{1}{t-z} = \pm i \int_0^\infty e^{\mp is(t-z)} ds, \quad \text{Im } z \gtrless 0]$$

Bromwich Integrals at Infinity

We consider the Bromwich integral defined by

$$\mathscr{B}(g) = \frac{1}{2\pi i} \int_B e^{zt} g(z) dz \tag{2.11.17}$$

The path of integration lies in the complex plane and is defined by $\text{Re } z = x_0 > 0$, a constant, $-\infty < \text{Im } z = y < +\infty$, i.e. it is a straight line parallel to the imaginary axis.

We will assume (i) $g(z)$ is analytic for $\text{Re } z > 0$, (ii) $\lim_{|y| \to \infty} g(z) = 0$, $x > 0$; (iii) g is absolutely integrable on B. These assumptions guarantee the existence of $\mathscr{B}(g)$ and also permit us to vary the path of integration. A major property of $\mathscr{B}(g)$ is that under sufficient assumptions it is the inverse of the Laplace transform, i.e. $\mathscr{L}(\mathscr{B}(g)) = g$. (See Doetsch, "Handbuch der Laplace Transform", Birkhauser Verlag, Basel, 1955.) In this same vein

Exercise 56. Prove that if

$$\lim_{x \to \infty} \int_B |g(x+iy)|\, dy = 0$$

then

$$\mathscr{B}(g) = 0, \quad t < 0.$$

We are interested in $\mathscr{B}(g)$ for $t \to \infty$. This clearly depends on how far to the left we may push the path B. If the path may be moved to a Bromwich path on which $\operatorname{Re} z = x_1$ then clearly

$$\mathscr{B}(g) = O(e^{x_1 t})$$

From this it follows that under our assumptions on g, $\mathscr{B}(g)$ decays in time less rapidly than an exponential.

In order to obtain more detail, it is necessary to further specify $g(z)$. A typical situation is that g satisfies the above assumptions (i) - (iii) for $\operatorname{Re} z > x_1$ where say $x_1 < 0$, except at one point z_0, $\operatorname{Re} z_0 > x_1$, which is a branch point of g. Under these circumstances we may distort the path of integration, B, into $R^- + L^- + L^+ + R^+$ indicated on the figure.

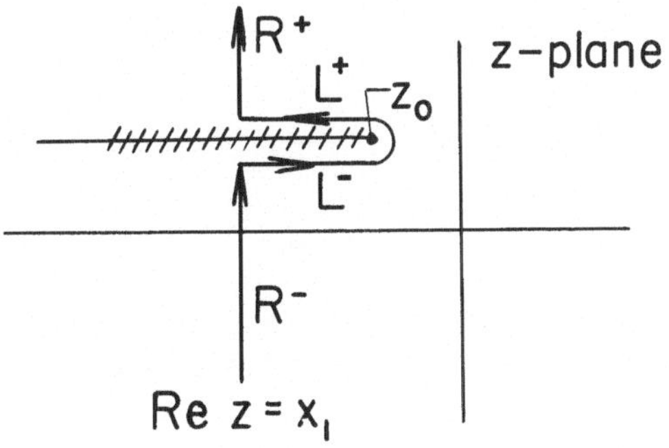

Then since

$$\frac{1}{2\pi i}\int_{R^-+R^+} g(z)e^{zt}dz = O(e^{x_1t})$$

we can write

$$\mathscr{B}(g) = \frac{1}{2\pi i}\int_{L^-+L^+} g(z)e^{zt}dz + O(e^{x_1t})$$

The branch cut at z_o is placed as indicated in the figure. Also we write $g^{\pm}$ for the evaluation of $g(z)$ at the upper and lower sides of the cut, and set

$$[g] = g^+ - g^-$$

Then from elementary changes of variable we obtain

$$\frac{1}{2\pi i}\int_{L^-+L^+} g(z)e^{zt}dz = -\frac{e^{z_ot}}{2\pi i}\int_0^{|x_1-x_o|} e^{-pt}[g(z_o-p)]dp$$

This is in the form of a Laplace transform, so that the expansion of $[g(z_o-p)]$ in the neighborhood of $p = 0$ furnishes the AD of $\mathscr{B}(g)$ for $t \to \infty$.

In the neighborhood of a branch point we can generally write

(2.11.18) $$g \sim \sum_{\nu=0} c_\nu(z-z_o)^{\delta_\nu-1}$$

with

$$\operatorname{Re}\delta_\nu > \operatorname{Re}\delta_{\nu-1}$$

For the moment we assume that $\operatorname{Re}\delta_\nu > 0$. Setting $z = z_o-p$

$$g \sim \sum_{\nu=0} c_\nu(-p)^{\delta_\nu-1}$$

and hence

$$g^{\pm} \sim \sum_{\nu=0} c_\nu (pe^{\pm i\pi})^{\delta_\nu - 1}$$

so that

$$[g] \sim -2i \sum_{\nu=0} c_\nu p^{\delta_\nu - 1} \sin(\delta_\nu \pi)$$

Finally from Watson's lemma we have

$$\mathscr{B}(g) \sim \frac{e^{z_0 t}}{\pi} \sum_{\nu=0} c_\nu \frac{(\delta_\nu - 1)!}{t^{\delta_\nu}} \sin(\delta_\nu \pi) \tag{2.11.19}$$

The condition that $\operatorname{Re} \delta_\nu > 0$ is easily removed. For suppose

$$g = \sum_{\mu=0}^{\mu_0} d_\mu (z-z_0)^{\varepsilon_\mu} + \hat{g}$$

where $\hat{g}$ has an expansion of the type (2.11.18) and

$$\operatorname{Re} \varepsilon_{\mu+1} > \operatorname{Re} \varepsilon_\mu$$

and

$$\operatorname{Re} \mu_0 \leq -1$$

We may therefore consider

$$\frac{1}{2\pi i} \int_B e^{zt} \sum_{\mu=0}^{\mu_0} d_\mu (z-z_0)^{\varepsilon_\mu} dz$$

If ε_μ is a negative integer the evaluation follows from residues. Otherwise we note

$$\int_B e^{zt}(z-z_o)^{\varepsilon_\mu} dz = \int_B e^{zt} \frac{d}{dz} \frac{(z-z_o)^{\varepsilon_\mu+1}}{\varepsilon_\mu+1} dz = -t\int_B e^{zt} \frac{(z-z_o)^{\varepsilon_\mu+1}}{\varepsilon_\mu+1} dz$$

This can be continued k times until $0 > \text{Re}\,(\varepsilon_\mu+k) > -1$. At this point the path B can be distorted into $R^- + L^- + L^+ + R^+$ and the previous treatment applied.

Exercise 57. Find three terms in the AD for $t \to \infty$ of $\mathscr{B}(g)$ for

(a) $g = (z+1)^{-5/2}$

(b) $g = (z+1)^{-1}(z+i+1)^{-1/2}(z-i+1)^{-1/3}$

(c) $g = e^{-z^{1/2}} z^{-2}$

Bromwich Integrals at the Origin

From the conditions on $g(z)$ we can write

$$\mathscr{B}(g) = \frac{1}{2\pi i} \int_{-\infty}^{\infty} e^{iyt} g(iy)\,dy$$

and

$$g = O(z^{-\mu})$$

$$\text{Re}\,\mu > 0$$

Hence the study of Bromwich integrals for $t \sim 0$ reduces to the comparable study for Fourier integrals and we refer simply to (2.11.16).

CHAPTER 3

LINEAR ORDINARY DIFFERENTIAL EQUATIONS

3.0. Introduction.

We will often write a derivative d/dx, as d_x and also $d_x^k u = u^{(k)}$. Thus a single n^{th} order linear inhomogeneous ordinary differential equation may be written as

$$u^{(n)} + a_1 u^{(n-1)} + \cdots a_n u = g(x) \tag{3.0.1}$$

where the coefficients a_i depend on x. (3.0.1) may also be written as a system of first order ordinary differential equation by defining new dependent variables

$$w_1 = u^{(n-1)}, \quad w_2 = u^{(n-2)}, \ldots, w_n = u.$$

Then since

$$w_{k-1} = \frac{dw_k}{dx}, \qquad k = 2, \ldots, n \tag{3.0.2}$$

we write (3.0.1) as

$$\frac{d}{dx}\begin{bmatrix} w_1 \\ \cdot \\ \cdot \\ \cdot \\ w_n \end{bmatrix} + \begin{bmatrix} a_1 & a_2, \ldots, a_n \\ -1 & 0 \ldots 0 \\ 0 & -1 \ldots 0 \\ \cdot & \cdot \quad \cdot \quad \cdot \\ 0 & \cdot \; \cdot \; -1 \quad 0 \end{bmatrix} \begin{bmatrix} w_1 \\ \cdot \\ \cdot \\ \cdot \\ w_n \end{bmatrix} = \begin{bmatrix} g \\ 0 \\ \cdot \\ \cdot \\ 0 \end{bmatrix} \tag{3.0.3}$$

or symbolically

$$\frac{d}{dx} \underset{\sim}{w} + \underset{\approx}{a}\underset{\sim}{w} = \underset{\sim}{g}.$$

Conversely we can consider a system of n first equations

$$\frac{d}{dx} w = \underset{\sim}{A}\underset{\sim}{w} + \underset{\sim}{f}(x) \qquad (3.0.4)$$
$$\underset{\sim}{A} = \underset{\sim}{A}(x)$$

and

<u>Exercise 58.</u> Demonstrate that by rational operations and differentiation any component of $\underset{\sim}{w}$ in (3.0.4) satisfies an ordinary differential equation of at most n^{th} order.

It is clear that we may treat systems of ordinary differential equations of any order either as a system of first order ordinary differential equations or as a single ordinary differential equation of some high order. Having demonstrated this equivalence it must also be mentioned that (3.0.1) and (3.0.4) each have some aspects that recommend study of it. From the point of view of clarity and theory it is more advantageous to consider the system (3.0.4). Even in describing the construction of a solution, (3.0.4) is found to be simpler to deal with. On the other hand, in terms of actual labor (3.0.1) is invariably simpler to solve. Especially since certain "tricks" for solving (3.0.1) have no counterpart for the system, (3.0.4). Moreover, unless care is taken (3.0.1) and (3.0.4) each produce spurious solutions when applied to the other. Therefore, in spite of the loss efficiency we will consider both the single equation and the system.

In keeping with the spirit of the earlier chapters, we shall be more interested in constructing approximate solutions than in the demonstration of the existence and uniqueness of solutions. We, therefore, state without proof: (see e.g., Coddington and Levinson, "Theory of Ordinary Differential Equations" [10]).

<u>Theorem 301.</u> For $\underset{\sim}{A}(z), \underset{\sim}{f}(z)$ analytic in the complex variable z in a region R, the equation

$$\frac{d\underset{\sim}{w}}{dz} = \underset{\sim}{A}(z)\underset{\sim}{w} + \underset{\sim}{f}(z) \qquad (3.0.5)$$

possesses a unique analytic solution for $z \in R$, taking on the prescribed data

$$(3.0.6)\qquad \underset{\sim}{w}(a) = \underset{\sim}{\alpha}, \qquad a \in R.$$

And from this,

<u>Theorem 302.</u> For $a_1(z),\ldots,a_n(z),f(z)$ analytic in R

$$(3.0.7)\qquad u^{(n)} + a_1(z)u^{(n-1)} + \cdots + a_n(z)u = f(z)$$

possesses a unique analytic solution in R taking on the prescribed data, $u^{(i)}(a) = \beta_i$, $i = 0,\ldots,n-1$, $a \in R$.

As an immediate application we solve (3.0.5) for analytic $\underset{\sim}{A}$ and $\underset{\sim}{f}$ subject to the data (3.0.6). For from Theorem 301 we can write the solution in the form

$$(3.0.8)\qquad \underset{\sim}{w}(z) = \sum_{i=0}^{\infty} \underset{\sim}{\alpha}_i(z-a)^i$$

with $\underset{\sim}{\alpha}_0 = \underset{\sim}{\alpha}$. From the analyticity of $\underset{\sim}{A}$ and $\underset{\sim}{f}$ we can also write,

$$\underset{\sim}{A} = \sum_{i=0}^{\infty} \underset{\sim}{A}_i(z-a)^i$$

$$\underset{\sim}{f} = \sum_{i=0}^{\infty} \underset{\sim}{f}_i(z-a)^i.$$

Substituting into (3.0.4) and equating powers we obtain

$$(3.0.9)\qquad \underset{\sim}{\alpha}_{i+1} = \frac{1}{i+1} \sum_{\substack{m+n=i\\ m,n\geq 0}} \underset{\sim}{A}_m\underset{\sim}{\alpha}_n + \underset{\sim}{f}_i \ , \qquad i \geq 1.$$

In a similar way we may construct a solution to the analytic initial value problem of equation (3.0.7).

It is clear that the above construction has a circle of convergence at least up to the first singularity of $\underset{\sim}{f}$ or $\underset{\sim}{A}$. That it may be larger is clear from the problem

$$\frac{dy}{dz} = \frac{y}{z-1}, \qquad y(0) = -1$$

which has the solution, $y = z - 1$.

The a priori knowledge of the possible singularities of the solution to an ordinary differential equation, is strictly a property of linear equations. For non-linear equations the location of a singular point is mobile and depends on the data of the problem. As an example consider,

$$\frac{dy}{dz} = y^2$$

which has the general solution

$$y = -\frac{1}{z+c}$$

with c determined by the data.

A nonlinear equation for which some a priori discussion of the singularities may be given is Riccati's equation

$$\frac{dy}{dz} = \alpha_0(z) + \alpha_1(z)y + \alpha_2(z)y^2 \tag{3.0.10}$$

where the α's are holomorphic in some region R say. Under the transformation

$$y = -\frac{1}{\alpha_2(z)}\frac{w'}{w}$$

this becomes,

$$\begin{gathered} w''(z) + a(z)w' + b(z)w = 0 \\ a(z) = -\frac{\alpha_2'}{\alpha_2} - \alpha_1, \quad b = \alpha_0\alpha_2. \end{gathered} \tag{3.0.11}$$

Exercise 59. (a) Discuss the singularities of (3.0.10) by writing the solution of (3.0.10) in terms of the solutions of (3.0.11).

(b) Give a transformation which reduces

$$\frac{d^2y}{dz^2} + a(z)\frac{dy}{dz} + b(z)y = 0 \tag{3.0.12}$$

to a first order equation.

Exercise 60. The Airy ordinary differential equation is

$$w'' - zw = 0 \tag{3.0.13}$$

(a) Reduce this to a system of first order ordinary differential equation.

(b) Find two (series) solutions of this equation. What are their circles of convergence?

Exercise 61. The parabolic cylinder ordinary differential equation is

$$\frac{d^2y}{dz^2} + (az^2+bz+c)y = 0 \tag{3.0.14}$$

a,b,c constants.

(a) Reduce this to "standard form"

$$y'' + \left(\frac{z^2}{4} - \alpha\right)y = 0. \tag{3.0.15}$$

(b) Find two different solutions.

(c) Write the ordinary differential equation as an equivalent first order system.

Before proceeding further it is necessary to consider some topics from matrix theory.

3.1. Some Topics in Matrix Analysis.

The purpose of this and the next section is to collect together those ideas and results from linear algebra which will have a direct bearing on our study of ordinary differential equations. For the most part the proofs will be assumed. (For proofs, see e.g., Halmos, "Finite Dimensional Vector Spaces", Princeton.)

A vector in n-space is written as $\underset{\sim}{v} = \begin{bmatrix} v_1 \\ \vdots \\ v_n \end{bmatrix}$ and its transpose as $\underset{\sim}{v}^T = [v_1,\ldots,v_n]$. Unless an ambiguity can arise we shall use $\underset{\sim}{v}$ to denote both the vector and its transpose. In general the entries of $\underset{\sim}{v}$ will be complex valued functions of the complex variable z and we will write $\underset{\sim}{v}(z)$.

The inner product between $\underset{\sim}{w}$ and $\underset{\sim}{v}$ is defined as

$$(3.1.1) \qquad (\underset{\sim}{w},\underset{\sim}{v}) = \sum_{i=1}^{n} w_i^* v_i$$

where the asterisk denotes the complex conjugate. From this we have

$$(\underset{\sim}{w},\underset{\sim}{v}) = (\underset{\sim}{v},w)^*$$

and

$$(a_1\underset{\sim}{w}+a_2\underset{\sim}{u},\underset{\sim}{v}) = a_1^*(\underset{\sim}{w},\underset{\sim}{v}) + a_2^*(\underset{\sim}{u},\underset{\sim}{v})$$

and

$$(\underset{\sim}{w},\underset{\sim}{w}) = 0 \Rightarrow \underset{\sim}{w} = 0.$$

(Note that (3.1.1) is not a function space inner product.)

As is easily demonstrated a linear operator in n-space is a matrix, say $\underset{\sim}{A}$.

Generally, these will be denoted by capitals and vectors by lower case letters. The determinant of a matrix will be represented by

$$\det \underset{\sim}{A} = |\underset{\sim}{A}|.$$

We write the transpose of $\underset{\sim}{A}$ as $\underset{\sim}{A}^t, (\underset{\sim}{A}^t)_{ij} = (\underset{\sim}{A})_{ji}$, and the complex conjugate transpose or adjoint by $\underset{\sim}{A}^+ = \underset{\sim}{A}^{t*}$. Clearly

$$(\underset{\sim}{A}^+\underset{\sim}{x},\underset{\sim}{y}) = (\underset{\sim}{x},\underset{\sim}{A}\underset{\sim}{y})$$

for any $\underset{\sim}{x}$ and $\underset{\sim}{y}$. Also

$$|\underset{\sim}{A}| = |\underset{\sim}{A}^t|.$$

<u>Definition.</u> $\underset{\sim}{A}$ is said to be hermitian if $\underset{\sim}{A} = \underset{\sim}{A}^+$.

<u>Definition.</u> $\underset{\sim}{A}$ is said to be normal if $\underset{\sim}{A}^+\underset{\sim}{A} = \underset{\sim}{A}\underset{\sim}{A}^+$. Therefore, all hermitian matrices are normal.

Denote the elements of $\underset{\sim}{A}$ by $a_{ij} = (\underset{\sim}{A})_{ij}$, then the cofactor of $a_{ij}, \tilde{a}_{ij}$, is defined as $(-1)^{i+j}$ times the determinant of the $(n-1)$ square matrix gotten by eliminating the ith-row and j^{th}-column of $\underset{\sim}{A}$. We write

$$(\tilde{\underset{\sim}{A}})_{ij} = \tilde{a}_{ij}$$

and call this the cofactor matrix.

<u>Definition.</u> $\tilde{\underset{\sim}{A}}^t$ is said to be the classical adjoint of $\underset{\sim}{A}$.

If $|\underset{\sim}{A}| \neq 0$, one can directly verify that

$$\underset{\sim}{A}^{-1} = \frac{\tilde{\underset{\sim}{A}}^t}{|\underset{\sim}{A}|}$$

i.e., $\underset{\sim}{A}^{-1}\underset{\sim}{A} = \underset{\sim}{A}\underset{\sim}{A}^{-1} = I$. Also if $|\underset{\sim}{A}| \neq 0$ the unique solution of $\underset{\sim}{A}\underset{\sim}{x} = \underset{\sim}{f}$ is $\underset{\sim}{x} = \underset{\sim}{A}^{-1}\underset{\sim}{f}$.

A set of vectors $\underset{\sim}{v}_i$, $i = 1,\ldots,n$ is said to be linearly dependent if there exist constants β_i, $i = 1,\ldots,m$ not all zero, such that

$$\sum_{i=1}^{m} \beta_i \underset{\sim}{v}_i = 0.$$

If no such set of constants exist the vectors are said to be linearly independent.

If $|\underset{\sim}{A}| \neq 0$ the rows and columns of $\underset{\sim}{A}$ form a linearly independent set of vectors, for otherwise

$$\underset{\sim}{A}\underset{\sim}{\beta} = 0$$

and

$$\underset{\sim}{A}^t\underset{\sim}{\alpha} = 0$$

would have non-trivial solutions. The maximum number of rows (columns) is known as the row rank (column rank) of $\underset{\sim}{A}$. It is clear that the row and column rank of a square matrix is the same - and we speak simply of the rank.

A non-zero vector $\underset{\sim}{\alpha}$ such that

$$\underset{\sim}{A}\underset{\sim}{\alpha} = 0$$

is said to belong to the null space of $\underset{\sim}{A}$. (Clearly, $|\underset{\sim}{A}| = 0$ in such a case.) Denoting the rank of $\underset{\sim}{A}$ by m the dimension of the null space is $n - m$.

The main theorem concerning the solvability of

(3.1.2) $$\underset{\sim}{A}\underset{\sim}{x} = \underset{\sim}{y}$$

states that: This equation has a solution only if y is orthogonal to the null space of $\underset{\sim}{A}^+$ and if this is satisfied the solution is unique up to the null space of $\underset{\sim}{A}$, i.e., to any solution may be added elements of the null-space of $\underset{\sim}{A}$.

Applications to Ordinary Differential Equations.

We now pause in our review of matrix theory to apply some of the ideas to ordinary differential equations.

Consider the system of linear homogeneous first order equations

$$(3.1.3) \qquad \frac{d\underset{\sim}{w}}{dz} = \underset{\sim}{A}\underset{\sim}{w}.$$

Let $\underset{\sim}{w}_1(z),\ldots,\underset{\sim}{w}_m(z)$ represent m solutions. Then by linearity

$$\sum_{i=1}^{m} \alpha_i \underset{\sim}{w}_i(z)$$

defines a solution for any m (complex) constants α_i, $i = 1,\ldots,m$. Therefore, the solutions of (3.1.3) form a vector space over the complex numbers.

Theorem 311. The dimension of the solution space of equation (3.1.3) is n.

Proof. Suppose $m > n$ independent solutions $\underset{\sim}{w}_1(z),\ldots,\underset{\sim}{w}_m(z)$. But at any point $a \in R$ we can choose m constants, not all zero, so that

$$\sum_{i=1}^{m} b_i \underset{\sim}{w}_i(a) = 0.$$

But $\sum_{i=1}^{m} b_i \underset{\sim}{w}_i(z)$ is a solution to (3.1.3) hence by uniqueness

$$\sum_{i=1}^{m} b_i \underset{\sim}{w}_i(z) \equiv 0.$$

Hence the dimension of the solution space is $\leq n$. Next let $\underset{\sim}{w}_i(z)$ be the solution to (3.1.3) such that $\underset{\sim}{w}_i(a)$ has all entries except the ith equal to zero, and suppose these are linearly dependent. Then there exist b_i, $i = 1,\ldots,n$ not all zero such that

$$\sum_{i=1}^{n} b_i \underset{\sim}{w}_i(z) = 0.$$

But setting $z = a$, we obtain a contradiction.

Theorem 312. Any n independent solutions, $\underset{\sim}{w}_i(z)$, $i = 1,\ldots,n$, of (3.1.3) form a basis, i.e., any solution of (3.1.3) can be represented by the $\underset{\sim}{w}_i(z)$.

Proof. Let $\underset{\sim}{y}(z)$ be any non-trivial solution of (3.1.3). Then $\underset{\sim}{y}, \underset{\sim}{w}_1, \ldots, \underset{\sim}{w}_n$ are linearly dependent, i.e., there exist constants, not all zero, $-a_o, a_1, \ldots, a_n$ such that

$$a_o \underset{\sim}{y}(z) = \sum_{i=1}^{n} a_i \underset{\sim}{w}_i(z).$$

But $a_o \neq 0$, otherwise a contradiction and hence

$$\underset{\sim}{y} = \sum_{i=1}^{n} \frac{a_i}{a_o} \underset{\sim}{w}_i(z).$$

Definition. Any set of n linearly independent solutions $\underset{\sim}{w}_1, \ldots, \underset{\sim}{w}_n$ of (3.1.3) is called a fundamental system of solutions.

Definition. Let $\underset{\sim}{w}_i$, $i = 1,\ldots,n$ represent a fundamental system and write $(\underset{\sim}{w}_j)_i = W_{ij}$ then the matrix

$$(\underset{\sim}{W})_{ij} = W_{ij}$$

is called a fundamental matrix of (3.1.3). I.e., the columns of a fundamental matrix form a fundamental system.

Clearly,

(3.1.4) $$\frac{d}{dz}\underset{\sim}{W} = \underset{\sim\sim}{A}\underset{\sim}{W}.$$

Definition. Let $\hat{\underset{\sim}{w}}_i$ represent n solutions of (3.1.3) (not necessarily a fundamental system). Then the Wronskian matrix of this system is defined as

$$\omega = \det[(\hat{\underset{\sim}{w}}_j)_i].$$

Theorem 313. The Wronskian of a fundamental system vanishes nowhere. Whereas the Wronskian of any other n solutions vanishes identically.

Proof. If $\omega(z)$ vanishes at some point $z = a$, say, then there exist constants b_i, $i = 1,\dots,n$ such that

$$\sum_{i=1}^{n} b_i \underset{\sim}{w}_i(z) = 0.$$

But from uniqueness

$$\sum_{i=1}^{n} b_i \underset{\sim}{w}_i(z) \equiv 0.$$ Q.E.D.

For the homogeneous form of (3.0.1),

(3.1.5) $$u^{(n)} + a_1 u^{(n-1)} + \dots a_n u = 0$$

one defines the Wronskian as

$$\omega(z) = \begin{vmatrix} u_1(z) & \dots & u_n(z) \\ u_1^{(1)}(z) & \dots & u_n^{(1)}(z) \\ \vdots & & \vdots \\ u_1^{(n-1)}(z) & & u_n^{(n-1)}(z) \end{vmatrix}$$

where $u_1,\dots,u_n$ refer to n solutions of (3.1.5). Then one may

<u>Exercise 62</u>. Demonstrate that

$$\omega(z) = \omega(a)e^{-\int_a^z a_1(z')dz'} \tag{3.1.6}$$

and, in general, for (3.1.3) one can

<u>Exercise 63</u>. Show that

$$\omega(z) = \omega(a)e^{-\int_a^z \operatorname{tr} A(z')dz'}$$

where tr = trace. $(\operatorname{tr} \underset{\sim}{A} = \sum_{i=1}^{n} A_{ii})$

Various fundamental matrix solutions to (3.1.4) can be related in a simple way. Suppose $\underset{\sim}{Y}(z)$ is a fundamental solution of

$$\frac{d}{dz}\underset{\sim}{Y}(z) = \underset{\sim}{A}(z)\underset{\sim}{Y}(z).$$

Then the solution of

$$\begin{aligned} &\frac{d}{dz}\underset{\sim}{X} = \underset{\sim}{A}(z)\underset{\sim}{X}(z) \\ &\underset{\sim}{X}(a) = \underset{\sim}{X}^{o} \end{aligned} \tag{3.1.7}$$

is given by

$$\underset{\sim}{X}(z) = \underset{\sim}{Y}(z)\underset{\sim}{Y}^{-1}(a)\underset{\sim}{X}^{o}$$

and in the same way the vector equation

$$\begin{aligned} &\frac{d}{dz}\underset{\sim}{v}(z) = \underset{\sim}{A}(z)\underset{\sim}{v}(z) \\ &\underset{\sim}{v}(z = a) = \underset{\sim}{v}^{o} \end{aligned} \tag{3.1.8}$$

has the solution

$$\underset{\sim}{v}(z) = \underset{\sim}{X}(z)\underset{\sim}{X}^{-1}(a)\underset{\sim}{v}^{o} \tag{3.1.9}$$

where $\underset{\sim}{X}(z)$ refers to any fundamental matrix solution. In both cases the uniqueness theorem tells us that this is the same solution as obtained by any other method.

It is implicit in our discussion that by

$$\frac{d\underset{\sim}{A}(z)}{dz}$$

we mean the matrix whose elements are

$$\frac{dA_{ij}(z)}{dz} .$$

Therefore, since

$$(\underset{\sim}{A}\underset{\sim}{B})_{ij} = \sum_{k=1}^{n} A_{ik}B_{kj}$$

we have that

$$\frac{d}{dz}\,\underset{\sim}{A}(z)\underset{\sim}{B}(z) = \underset{\sim}{A}'(z)\underset{\sim}{B}(z) + \underset{\sim}{A}(z)\underset{\sim}{B}'(z).$$

Also by

$$\int \underset{\sim}{A}(z)dz$$

we shall mean the matrix whose elements are

$$\int A_{ij}(z)dz.$$

Using these conventions we consider the inhomogeneous matrix equation

$$(3.1.10) \qquad \frac{d}{dz}\underset{\sim}{X} = \underset{\sim\sim}{AX} + \underset{\sim}{F}(z)$$

with

$$\underset{\sim}{X}(a) = \underset{\sim}{X}^{o}.$$

To solve this problem we use the method known as variation of parameters. Let $\underset{\sim}{Y}(z)$ be any fundamental matrix solution of (3.1.4) and introduce a new unknown matrix $\underset{\sim}{Z}(z)$ through,

$$(3.1.11) \qquad \underset{\sim}{X}(z) = \underset{\sim}{Y}(z)\underset{\sim}{Z}(z).$$

Then substituting (3.1.11) into (3.1.10) we obtain

$$\underset{\sim}{Y}'(z)\underset{\sim}{Z}(z) + \underset{\sim}{Y}(z)\underset{\sim}{Z}'(z) = \underset{\sim\sim\sim}{AYZ} + F(z)$$

and hence

$$(3.1.12) \qquad \underset{\sim}{Z}'(z) = \underset{\sim}{Y}^{-1}(z)\underset{\sim}{F}(z).$$

From (3.1.11) the initial condition on $\underset{\sim}{Z}$ is

$$(3.1.13) \qquad \underset{\sim}{Z}(a) = \underset{\sim}{Y}^{-1}(a)\underset{\sim}{X}^{o}.$$

Therefore, integrating (3.1.12) and using the condition (3.1.13), we have

$$\underset{\sim}{Z}(z) = \underset{\sim}{Y}^{-1}(a)\underset{\sim}{X}^{o} + \int_a^z \underset{\sim}{Y}^{-1}(\zeta)\underset{\sim}{F}(\zeta)d\zeta$$

and finally,

$$\underset{\sim}{X}(z) = \underset{\sim}{Y}(z)\underset{\sim}{Y}^{-1}(a)\underset{\sim}{X}^{o} + \int_a^z \underset{\sim}{Y}(z)\underset{\sim}{Y}^{-1}(\zeta)\underset{\sim}{F}(\zeta)d\zeta. \tag{3.1.14}$$

Since we are dealing with analytic functions $\underset{\sim}{A}(z)$, (and hence, analytic $\underset{\sim}{Y}(z)$) $\underset{\sim}{F}(z)$ the path of integration in (3.1.14) is not important as long as it is in the common domain of analyticity.

For the vector problem

$$\begin{aligned} \frac{d}{dz}\underset{\sim}{v}(z) &= \underset{\sim}{A}(z)\underset{\sim}{v}(z) + \underset{\sim}{f}(z) \\ \underset{\sim}{v}(a) &= \underset{\sim}{v}^{o} \end{aligned} \tag{3.1.15}$$

we have

$$\underset{\sim}{v}(z) = \underset{\sim}{Y}(z)\underset{\sim}{Y}^{-1}(a)\underset{\sim}{v}^{o} + \int_a^z \underset{\sim}{Y}(z)\underset{\sim}{Y}^{-1}(\zeta)\underset{\sim}{f}(\zeta)d\zeta. \tag{3.1.16}$$

3.2. Matrix Theory - Continued.

The discussion of matrices is greatly facilitated by the introduction of eigenvalues and eigenvectors. An n-square matrix A is said to have an eigenvector $\underset{\sim}{v}$ and corresponding eigenvalue λ if

$$\underset{\sim}{A}\underset{\sim}{v} = \lambda\underset{\sim}{v}. \tag{3.2.1}$$

Hence for $\underset{\sim}{v}$ to be non-trivial we must have

$$P(\lambda) = |\underset{\sim}{A}-\lambda\underset{\sim}{I}| = 0. \tag{3.2.2}$$

The nth degree polynomial $P(\lambda)$ defined by (3.2.2) is referred to as the characteristic polynomial. Therefore, $P(\lambda)$ has n roots not necessarily all distinct. If λ_k is a root of multiplicity m then it is also said to be eigenvalue of multiplicity m of $\underset{\sim}{A}$.

If μ and λ are distinct eigenvalues of (3.2.1) and $\underset{\sim}{v}$ and $\underset{\sim}{w}$ their respective eigenvalues, they are linearly independent. For otherwise there would exist a $k \neq 0$ such that $\underset{\sim}{v} = k\underset{\sim}{w}$, then $\underset{\sim}{A}k\underset{\sim}{w} = \lambda k\underset{\sim}{w}$ and dividing by k, $\underset{\sim}{A}\underset{\sim}{w} = \lambda\underset{\sim}{w}$, which is a contradiction. Therefore, if all the eigenvalues $\lambda_1,\ldots,\lambda_n$ of $\underset{\sim}{A}$ are distinct the corresponding eigenvectors $\underset{\sim}{v}_1,\ldots,\underset{\sim}{v}_n$ are linearly independent. Denoting by $\underset{\sim}{V}$ the matrix whose columns are $\underset{\sim}{v}$, it follows that $\underset{\sim}{V}^{-1}$ exists. Also writing

$$\underset{\sim}{\Lambda} = \begin{bmatrix} \lambda_1 & & \bigcirc \\ & \ddots & \\ \bigcirc & & \lambda_n \end{bmatrix}$$

we have

$$\underset{\sim}{A}\underset{\sim}{V} = \underset{\sim}{V}\underset{\sim}{\Lambda}$$

from which it follows that

(3.2.3) $$\underset{\sim}{A} = \underset{\sim}{V}\underset{\sim}{\Lambda}\underset{\sim}{V}^{-1}$$

and

(3.2.4) $$\underset{\sim}{V}^{-1}\underset{\sim}{A}\underset{\sim}{V} = \underset{\sim}{\Lambda}.$$

Note that $|\underset{\sim}{A}| = |\underset{\sim}{\Lambda}|$.

When a transformation $\underset{\sim}{V}$ exists which diagonalizes a matrix in the way shown in (3.2.4) one says that $\underset{\sim}{A}$ is diagonalizable under similarity. Hence we have demonstrated that a matrix with distinct eigenvalues is diagonalizable under similarity. A more useful result in this direction is that a normal matrix is diagonalizable under similarity. Therefore, in particular, real symmetric and hermitian matrices are diagonalizable under similarity.

We note that a matrix diagonalized under similarity is composed of the eigenvalues of the matrix. For if there exists an $\underset{\sim}{S}$ for $\underset{\sim}{A}$ such that

$$\underset{\sim}{S}\underset{\sim}{A}\underset{\sim}{S}^{-1} = \underset{\sim}{D}$$

$\underset{\sim}{D}$ diagonal then also

$$\underset{\sim}{S}(\underset{\sim}{A}-\lambda\underset{\sim}{I})\underset{\sim}{S}^{-1} = \underset{\sim}{D} - \lambda\underset{\sim}{I}$$

and on taking the determinant of both sides we have the result.

Corresponding to each eigenvalue λ_i of a matrix, there is at least one and maybe many linearly independent eigenvectors. The eigenvectors corresponding to this eigenvalue clearly form a subspace and we denote this by $E(\lambda_i)$. Then if

$$\sum_i \dim E(\lambda_i) = n$$

the order of the matrix, it again is clear that the matrix in question is diagonalizable by similarity. In fact, diagonalization under similarity is equivalent to having n linearly independent eigenvectors. For we can write the eigenvalue equation in matrix form as $\underset{\sim}{A}\underset{\sim}{V} = \underset{\sim}{V}\underset{\sim}{\Lambda}$ and if $\underset{\sim}{V}$ is composed of n linearly independent eigenvectors, $\underset{\sim}{V}^{-1}$ exists and similarity follows. On the other hand, if $\underset{\sim}{A}$ is diagonalizable under similarity, $\underset{\sim}{S}^{-1}\underset{\sim}{A}\underset{\sim}{S} = \underset{\sim}{\Lambda}$. Then since $\underset{\sim}{S}^{-1}$ exists it has n linearly independent columns. Multiplying on the left by $\underset{\sim}{S}$ shows that the columns of $\underset{\sim}{S}$ are eigenvectors.

Consider the $k \times k$ matrix,

$$\underset{\sim}{L}_k(\lambda) = \begin{bmatrix} \lambda & 1 & 0 & \cdots & 0 \\ 0 & \lambda & 1 & \cdots & 0 \\ \cdot & \cdot & \cdot & \cdots & \cdot \\ 0 & 0 & & \cdots & 1 \\ 0 & 0 & & \cdots & \lambda \end{bmatrix}$$

with λ in the diagonal, 1 in the super diagonal and zero elsewhere. This is clearly an example of a matrix which is not diagonalizable under similarity. In fact, it has the single eigenvalue λ, with multiplicity k, and the space $E(\lambda)$ is of dimension one, and is spanned by $(1,0,\ldots,0)$.

Jordan Canonical Form.

One may demonstrate that for any matrix $\underset{\sim}{A}$, there exists a non-singular matrix $\underset{\sim}{T}$ such that

$$\underset{\sim}{T}^{-1}\underset{\sim}{A}\underset{\sim}{T} = \begin{bmatrix} \underset{\sim}{L}_{k_1}(\lambda_1) & 0 & \cdots & 0 \\ 0 & \underset{\sim}{L}_{k_2}(\lambda_2) & \cdots & 0 \\ \vdots & & \ddots & \vdots \\ 0 & 0 & \cdots & \underset{\sim}{L}_{k_r}(\lambda_r) \end{bmatrix} \tag{3.2.5}$$

where $\sum_{i=1}^{r} k_i = n$, and the λ_i (which of course are the eigenvalues) are not necessarily all distinct. The right hand side of (3.2.5) is called the Jordan canonical form of $\underset{\sim}{A}$.

For a 3×3 matrix with threefold degenerate eigenvalue λ, there are three possible Jordan forms,

$$\lambda\underset{\sim}{I}, \qquad \begin{pmatrix} \lambda & \\ 0 & L_2(\lambda) \end{pmatrix}, \qquad L_3(\lambda).$$

As an application of the Jordan form consider the homogeneous linear equation

$$(3.2.6) \qquad AX - XB = 0$$

with A an $n \times n$ matrix, B an $m \times m$ matrix and the unknown matrix, X having m columns and n rows.

Theorem 320. (3.2.6) has a non-trivial solution if and only if A and B have at least one common eigenvalue.

Proof. Place B in Jordan form

$$B = SJS^{-1}.$$

Equation (3.2.6) then has the equivalent form

$$(3.2.7) \qquad A(XS) = (XS)J.$$

Without loss of generality, we take the (1,1) entry of J be, λ, the common eigenvalue. Also writing $Av = \lambda v$ with $v^T = [v_1, \ldots, v_n]$, we see that

$$XS = \begin{bmatrix} 0 & 0 & \cdots & v_1 \\ 0 & 0 & \cdots & v_2 \\ \cdot & \cdot & \cdots & \\ 0 & 0 & \cdots & v_n \end{bmatrix}$$

is indeed a solution. To prove necessity suppose A and B do not have a common eigenvalue and $XS \neq 0$ in (3.2.7). Then starting with the first column and working toward the last we have a contradiction.

Next, it will be useful to introduce a norm of a matrix A, $\|A\|$. We define

$$\|\underset{\sim}{A}\| = \max_i \sum_j |(\underset{\sim}{A})_{ij}|. \tag{3.2.8}$$

Then from this definition it follows directly that (1) $\|\underset{\sim}{A}\| \geq 0$, (2) $\|\underset{\sim}{A}\| = 0$ if and only if $\underset{\sim}{A} = 0$, (3) $\|c\underset{\sim}{A}\| = |c| \, \|\underset{\sim}{A}\|$ for any complex c, (4) $\|\underset{\sim}{A}\| + \|\underset{\sim}{B}\| \geq \|\underset{\sim}{A}+\underset{\sim}{B}\|$. These conditions define a norm and therefore justify the definition (3.2.8).

Further one may

Exercise 64. Prove

(5) $\|\underset{\sim}{A}\underset{\sim}{B}\| \leq \|\underset{\sim}{A}\|\|\underset{\sim}{B}\|$

(b) $\lim_{n \to \infty} \underset{\sim}{A}_n = 0$, if and only if

$\lim_{n \to \infty} \|\underset{\sim}{A}_n\| = 0.$

(By the convergence of a matrix we mean the convergence of each entry of the matrix.)

Definitions other than (3.2.8) can be given for the norm of a matrix, the one given will be particularly useful. In the same vein we use instead of the usual vector norm (i.e., $(\underset{\sim}{v},\underset{\sim}{v})^{1/2}$) the norm

$$\|\underset{\sim}{v}\| = \max_j |v_j|.$$

With these definitions we can

Exercise 65. Demonstrate

$$\|\underset{\sim}{A}\underset{\sim}{v}\| \leq \|\underset{\sim}{A}\|\|\underset{\sim}{v}\|.$$

Let λ_j, $j = 1,2,\ldots$ represent the eigenvalues of $\underset{\sim}{A}$ and $\underset{\sim}{v}_j$ the corresponding eigenvectors. Then

$$|\lambda_j| \, \|\underset{\sim}{v}_j\| = \|\lambda_j\underset{\sim}{v}_j\| = \|\underset{\sim}{A}\underset{\sim}{v}_j\| \leq \|\underset{\sim}{A}\|\|\underset{\sim}{v}_j\|$$

and hence

Lemma. For the eigenvalues λ_j of $\underset{\sim}{A}$

$$\max_j |\lambda_j| \leq \|\underset{\sim}{A}\|.$$

The following theorem is important in dealing with functions of matrices.

<u>Theorem 321.</u> $\quad \sum_{r=0}^{\infty} \|\underset{\sim}{M}_r\| < \infty \Rightarrow \sum_{r=0}^{\infty} \underset{\sim}{M}_r < \infty.$

<u>Proof.</u> We show that partial sums are Cauchy sequences

$$\| \sum_{r=n}^{m} \underset{\sim}{M}_r \| \leq \sum_{r=n}^{m} \|\underset{\sim}{M}_r\|$$

which from the convergence of $\sum_{r=0}^{\infty} \|\underset{\sim}{M}_r\|$ can be made as small as we please by choosing m,n sufficiently large.

<u>Functions of Matrices.</u>

Since the monomial of a matrix $\underset{\sim}{A}, \underset{\sim}{A}^n$, is well-defined the polynomial of a matrix can also be defined. For if

$$p(x) = x^n + a_1 x^{n-1} + \cdots + a_n \quad (a_1, \ldots, a_n \text{ scalars})$$

we define the matrix

$$p(\underset{\sim}{A}) = \underset{\sim}{A}^n + a_1 \underset{\sim}{A}^{n-1} + \cdots + a_n \underset{\sim}{I}. \tag{3.2.9}$$

For the definition of a function of a matrix under more general circumstances, let us first consider those $\underset{\sim}{A}$ that are diagonalizable under similarity

$$\underset{\sim}{A} = \underset{\sim}{S}\underset{\sim}{D}\underset{\sim}{S}^{-1} \tag{3.2.10}$$

where $\underset{\sim}{D}$ is the diagonal matrix of eigenvalues. Placing this into (3.2.9)

$$p(\underset{\sim}{A}) = \underset{\sim}{S}(\underset{\sim}{D}^n+a_1\underset{\sim}{D}^{n-1}+\cdots+a_n\underset{\sim}{I})\underset{\sim}{S}^{-1}$$

$$= \underset{\sim}{S}p(\underset{\sim}{D})\underset{\sim}{S}^{-1} = \underset{\sim}{S}\begin{bmatrix} p(\lambda_1) & 0 & \cdots & 0 \\ 0 & p(\lambda_2) & \cdots & 0 \\ \cdots & \cdots & \cdots & \cdots \\ 0 & 0 & \cdots & p(\lambda_n) \end{bmatrix}\underset{\sim}{S}^{-1}$$

where $\lambda_1,\ldots,\lambda_n$ are the eigenvalues of $\underset{\sim}{A}$.

Hence for any continuous function $f(x)$ and $\underset{\sim}{A}$ such that $\underset{\sim}{A} = \underset{\sim}{S}\underset{\sim}{D}\underset{\sim}{S}^{-1}$ we can define the matrix,

$$f(\underset{\sim}{A}) = \underset{\sim}{S}f(\underset{\sim}{D})\underset{\sim}{S}^{-1}$$

where

$$(3.2.12) \qquad f(D) = \begin{bmatrix} f(\lambda_1) & 0 & \cdots & 0 \\ \cdots & \cdots & \cdots & \cdots \\ 0 & 0 & \cdots & f(\lambda_n) \end{bmatrix}$$

We pause in this discussion to point out that if $P(\lambda)$ is the characteristic polynomial (3.2.2), and λ_i is an eigenvalue of $\underset{\sim}{A}$, then $P(\lambda_i) = 0$, hence for $\underset{\sim}{A}$ such that (3.2.10) holds, we have shown by virtue of (3.2.11), that

$$(3.2.13) \qquad P(\underset{\sim}{A}) = 0.$$

This, however, can be shown to be true independently of property (3.2.10).

<u>Theorem (Cayley-Hamilton)</u>. A square matrix satisfies its characteristic polynomial. [An almost trivial proof of this is given below.] This very useful result will arise again shortly. In passing we point out that multiplying (3.2.13) by $\underset{\sim}{A}^{-1}$ (if it exists) gives a representation of $\underset{\sim}{A}^{-1}$ as a polynomial of degree $n - 1$ in $\underset{\sim}{A}$.

We can arrive at another definition of a function of a matrix as follows. Consider $g(z)$ analytic in the neighborhood of the origin, and such that

$$g(z) = \sum_{n=0}^{\infty} a_n z^n < \infty$$

for $|z| < R$.

Then for any matrix $\underset{\sim}{A}$ such that $\|\underset{\sim}{A}\| < R$, it follows by comparison and Theorem 321 that

$$g(\underset{\sim}{A}) = \sum_{n=0}^{\infty} a_n \underset{\sim}{A}^n < \infty. \tag{3.2.14}$$

To see that this incorporates the previous definition, let $\underset{\sim}{A}$ satisfy (3.2.10), then

$$g(\underset{\sim}{A}) = \underset{\sim}{S}\left(\sum_{n=0}^{\infty} a_n \underset{\sim}{D}^n\right)\underset{\sim}{S}^{-1} = \underset{\sim}{S}g(\underset{\sim}{D})\underset{\sim}{S}^{-1}.$$

The matrix $g(\underset{\sim}{D})$ (as a series) exists since $\max |\lambda_i| \leq \|\underset{\sim}{A}\| < R$, where λ_i represents, as usual, the eigenvalues of $\underset{\sim}{A}$.

As a last step in this progression of defining a function of a matrix, we wish to eliminate the requirement that the radius of convergence, R, of $g(z)$ be larger than the spectral radius $\max\limits_i |\lambda_i|$ of $\underset{\sim}{A}$. To do this let us first introduce the resolvent operator of $\underset{\sim}{A}$,

$$\underset{\sim}{R} = \frac{1}{\underset{\sim}{A}-z\underset{\sim}{I}} . \tag{3.2.15}$$

The z-plane with $\lambda_1,\ldots,\lambda_k$ deleted is called the resolvent set of $\underset{\sim}{R}$. By writing

$$\underset{\sim}{R}(z) = \frac{\underset{\sim}{B}(z)}{P(z)} \tag{3.2.16}$$

where $\underset{\sim}{B}$ is the classical adjoint of $(\underset{\sim}{A}-z\underset{\sim}{I})$ (see definition, page 193) and $P(z)$ the characteristic polynomial (3.2.2), we see that $R(z)$ exists for z belonging to the resolvent set and that it is singular at an eigenvalue of $\underset{\sim}{A}$.

Next, we consider $\underset{\sim}{R}(z)$ for $|z| > \|\underset{\sim}{A}\|$, then we can write

$$\frac{1}{z\underset{\sim}{I}-\underset{\sim}{A}} = \frac{1}{z} \sum_{n=0}^{\infty} \frac{\underset{\sim}{A}^n}{z^n} < \infty.$$

The convergence is uniform and we integrate over a closed contour on which $|z| = c > \|\underset{\sim}{A}\|$, say. The result is

$$\frac{1}{2\pi i} \oint \frac{dz}{z\underset{\sim}{I}-\underset{\sim}{A}} = \underset{\sim}{I}. \tag{3.2.17}$$

More generally if $g(z)$ is analytic for $|z| \leq c$, we obtain

$$\frac{1}{2\pi i} \oint_{|z|=c} \frac{g(z)}{z-\underset{\sim}{A}} dz = \sum_{n=0}^{\infty} \frac{g^n(0)}{n!} \underset{\sim}{A}^n \tag{3.2.18}$$

which is the same as (3.2.14). However, in both (3.2.17) and (3.2.18) the integral does not depend on the particular contour. In fact, by Cauchy's theorem the contour may be distorted into a small circle around each of the eigenvalues of $\underset{\sim}{A}$. This then leads us to our final definition.

<u>Definition (Dunford-Taylor Integral)</u>. Let $g(z)$ be analytic in the neighborhood of each of the eigenvalues of a matrix $\underset{\sim}{A}$. Then the matrix $g(\underset{\sim}{A})$ is defined as

$$g(\underset{\sim}{A}) = \frac{1}{2\pi i} \oint_{\Gamma} \frac{g(z)dz}{z-\underset{\sim}{A}} \tag{3.2.19}$$

where Γ is any contour (or collection of contours) enclosing the eigenvalues of $\underset{\sim}{A}$ but not the singularities of $g(z)$.

If $\underset{\sim}{A}$ is diagonalizable under similarity $\underset{\sim}{A} = \underset{\sim}{S}\underset{\sim}{D}\underset{\sim}{S}^{-1}$, the Dunford-Taylor Integral becomes

$$\begin{aligned} g(\underset{\sim}{A}) &= \frac{1}{2\pi i} \oint_{\Gamma} \frac{g(z)dz}{\underset{\sim}{S}(z-\underset{\sim}{D})\underset{\sim}{S}^{-1}} = \underset{\sim}{S}\Big(\frac{1}{2\pi i} \oint_{\Gamma} \frac{g(z)dz}{z-\underset{\sim}{D}}\Big)\underset{\sim}{S}^{-1} \\ &= \underset{\sim}{S}g(\underset{\sim}{D})\underset{\sim}{S}^{-1}. \end{aligned}$$

Therefore, this and the above remarks show that the definition (3.2.19) is compatible with the previous definitions. It is important to note that it is not more general than (3.2.12). For when $\underset{\sim}{A}$ is diagonalizable under similarity, definition (3.2.12) applies even to functions which are only continuous. By contrast (3.2.19) requires analyticity, but of course it is defined for all matrices.

Construction of a Function of a Matrix.

Next, we seek rules for the explicit evaluation of the function of a matrix. Let z be a point of the resolvent set. Then considering the resolvent of the $n \times n$ matrix $\underset{\sim}{A}$ given in the form (3.2.16), we remark that $P(z)$ is of degree n and $\underset{\sim}{B}(z)$, the classical adjoint, has polynomial entries which are of at most degree $n - 1$. We may, therefore, write for the classical adjoint of $\underset{\sim}{A} - z\underset{\sim}{I}$,

$$\underset{\sim}{B} = z^{n-1}\underset{\sim}{B}_1 + z^{n-2}\underset{\sim}{B}_2 + \cdots + \underset{\sim}{B}_n.$$

Also we expand the characteristic polynomial

$$P(z) = z^n + P_1 z^{n-1} + \cdots + P_n$$

and rewrite (3.2.16) as

$$(3.2.20) \qquad (z^n+P_1z^{n-1}+\cdots+P_n)\underset{\sim}{1} = (\underset{\sim}{A}-z\underset{\sim}{I})(z^{n-1}\underset{\sim}{B}_1+\cdots+\underset{\sim}{B}_n).$$

Then equating coefficients of z^k, $k = 0,\ldots,n$, we obtain

$$\underset{\sim}{A}\underset{\sim}{B}_j - \underset{\sim}{B}_{j+1} = P_j\underset{\sim}{1}$$

with $\underset{\sim}{B}_0 = 0$ and $P_0 = 1$. We note from this construction that the $\underset{\sim}{B}_i$ can be successively computed and that in general $\underset{\sim}{B}_j$ is a polynomial of degree $j - 1$ in the matrix $\underset{\sim}{A}$. This then allows the alternate representation of $\underset{\sim}{B}$,

$$\underset{\sim}{B} = b_0 \underset{\sim}{1} + b_1 \underset{\sim}{A} + \cdots + b_{n-1} \underset{\sim}{A}^{n-1}$$

where the b_i are polynomials of degree at most $i - 1$ in z. [We note in passing that (3.2.20) remains valid if z is replaced by any matrix which commutes with $\underset{\sim}{A}$. In particular, if $\underset{\sim}{A}$ is substituted for z we prove the Cayley-Hamilton theorem. An even simpler proof follows from (3.2.19) by taking $g(z) = P(z)$.]

Next suppose $P(z)$ and $\underset{\sim}{B}(z)$ have a common factor,

$$P(z) = \alpha(z)\hat{P}(z)$$

$$\underset{\sim}{B}(z) = \alpha(z)\hat{\underset{\sim}{B}}(z).$$

We then have instead of (3.2.19)

$$(3.2.21) \qquad g(\underset{\sim}{A}) = \frac{1}{2\pi i} \oint_\Gamma \frac{g(z)\hat{\underset{\sim}{B}}(z)dz}{\hat{P}(z)} .$$

In particular

$$\hat{P}(\underset{\sim}{A}) = 0$$

$\hat{P}(z)$ is known as the minimal polynomial, and

$$\deg \hat{P} = m \leq n.$$

Using the same construction as above (3.2.20) we can write

$$\hat{\underset{\sim}{B}} = \hat{b}_0 \underset{\sim}{1} + \hat{b}_1 \underset{\sim}{A} + \cdots + \hat{b}_{m-1} \underset{\sim}{A}^{m-1}.$$

Substituting this into (3.2.21) we have

$$g(\underset{\sim}{A}) = \sum_{k=0}^{m-1} \frac{\underset{\sim}{A}^k}{2\pi i} \oint_\Gamma \frac{g(z)\hat{b}_k(z)dz}{\hat{P}(z)} .$$

Hence whatever analytic function $g(z)$ we consider (provided it is analytic in neighborhood of the eigenvalues of $\underset{\sim}{A}$), $g(\underset{\sim}{A})$ is expressible as a polynomial in $\underset{\sim}{A}$ of degree $m - 1$, where m is the degree of the minimum polynomial.

As an application, suppose Q is a polynomial, then by division we can find polynomials $s(x)$ and $r(x)$, with $\deg r < n$ such that

$$Q(x) = s(x)P(x) + r(x)$$

and therefore

$$Q(\underset{\sim}{A}) = r(\underset{\sim}{A}).$$

To proceed more generally let $g(z)$ be an appropriate analytic function and let us for the moment consider $\underset{\sim}{A}$ to be simple (i.e., distinct eigenvalues). Denote the eigenvalues of $\underset{\sim}{A}$ by $\lambda_1,\dots,\lambda_n$. Next construct a polynomial $r(x)$

$$r(x) = a_1 z^{n-1} + \cdots + a_n$$

according to the n-conditions

$$r(\lambda_i) = g(\lambda_i), \quad i = 1,\dots,n.$$

This results in n linear equations for the solution of $a_1,\dots,a_n$. Consider

$$\frac{g(z)-r(z)}{P(z)} = q(z). \tag{3.2.22}$$

Clearly, $q(z)$ is analytic in the neighborhood of each λ_i. Then writing

$$g(z) = P(z)q(z) + r(z) \tag{3.2.23}$$

and substituting into (3.2.19) demonstrates that

(3.2.24) $$g(\underset{\sim}{A}) = r(\underset{\sim}{A}).$$

When the eigenvalues of $\underset{\sim}{A}$ are not simple the minimal polynomial and the characteristic polynomial may be different. Suppose the minimal polynomial has the form,

$$\hat{P}(z) = (z-\lambda_1)^{k_1}(z-\lambda_2)^{k_2} \cdots (z-\lambda_\mu)^{k_\mu}$$

with

$$\sum_{i=1}^{\mu} k_i = m \leq n.$$

Then for a $g(z)$ we construct an $\hat{r}(z)$

$$\hat{r}(z) = \hat{a}_1 z^{m-1} + \cdots + \hat{a}_m$$

by

(3.2.25) $$\left.\frac{d^p}{dz^p}\hat{P}(z)\right|_{z=\lambda_i} = \left.\frac{d^p}{dz^p} r(z)\right|_{z=\lambda_i} \qquad p = 0,\ldots,k_i - 1$$

for $i = 1,\ldots,\mu$. Then as above

(3.2.26) $$\frac{g(z)-\hat{r}(z)}{\hat{P}(z)} = \hat{q}(z)$$

is analytic in the neighborhood of the eigenvalues and it follows that

(3.2.27) $$g(\underset{\sim}{A}) = \hat{r}(\underset{\sim}{A}).$$

In actual practice finding the minimal polynomial can be tedious and it is more expeditious to directly apply the above procedure to the characteristic polynomial, P(z).

<u>Example.</u> Evaluate

$$\exp\left(t\begin{bmatrix} 6 & -1 \\ 3 & 2 \end{bmatrix}\right)$$

where t is to be regarded as a parameter. First, we find the eigenvalues

$$\det\begin{bmatrix} 6-\lambda & -1 \\ 3 & 2-\lambda \end{bmatrix} = 0$$

$\lambda = 3,5$. Next we determine

$$r(z) = \alpha + \beta z$$

by

$$e^{tz}\Big|_{z=3,5} = (\alpha+\beta z)\Big|_{z=3,5}$$

or

$$\begin{aligned} e^{3t} &= \alpha + \beta 3 \\ e^{5t} &= \alpha + \beta 5 \\ \alpha &= (5e^{3t}-3e^{5t})/2 \\ \beta &= (e^{5t}-e^{3t})/2. \end{aligned}$$

Therefore

$$e^{tA} = \alpha \underset{\sim}{1} + \beta \underset{\sim}{A}$$

$$= \frac{1}{2}\begin{pmatrix} 3e^{5t} - e^{3t} & e^{3t} - e^{5t} \\ 3e^{5t} - 3e^{3t} & 3e^{3t} - e^{5t} \end{pmatrix}$$

<u>Exercise 66</u>. Evaluate

(1) $$e^{\begin{pmatrix} 0 & 0 & 0 \\ 0 & 0 & 1 \\ 0 & 0 & 0 \end{pmatrix} t}$$

(2) $$\sin \begin{pmatrix} 1 & 1 \\ 0 & 2 \end{pmatrix} t.$$

<u>Exercise 67</u>. Evaluate

$$\begin{pmatrix} a & b \\ c & d \end{pmatrix}^n, \quad \begin{pmatrix} \cos\theta, & \sin\theta \\ -\sin\theta, & \cos\theta \end{pmatrix}^n$$

<u>Exercise 68</u>. Evaluate

$$\exp\left[t \begin{pmatrix} 1 & 2 & 3 \\ -1 & 3 & 1 \\ 1 & 0 & 2 \end{pmatrix} \right]$$

In what follows the three matrix functions

$$e^{\underset{\sim}{A}}, \quad x^{\underset{\sim}{A}}, \quad \ln \underset{\sim}{A}$$

figure in an important way.

The first of these we have discussed at length.

The second by the Dunford-Taylor integral is

$$x^{\underset{\sim}{A}} = \frac{1}{2\pi i} \oint_{\Gamma} \frac{x^z dz}{z-\underset{\sim}{A}} =$$

$$= \frac{1}{2\pi i} \oint_{\Gamma} \frac{e^{z \ln x}}{z-\underset{\sim}{A}} dz = e^{\underset{\sim}{A} \ln x} .$$

The last relation, is naturally a property of $x^{\underset{\sim}{A}}$ which we want to retain. From it we see

$$(xe^{2\pi i})^{\underset{\sim}{A}} = e^{2\pi i\underset{\sim}{A}+\underset{\sim}{A} \ln x}$$

$$= e^{2\pi i\underset{\sim}{A}} x^{\underset{\sim}{A}} .$$

<u>Exercise 69</u>. Show if $\underset{\sim}{A}\underset{\sim}{B} = \underset{\sim}{B}\underset{\sim}{A}$, then

$$e^{\underset{\sim}{A}+\underset{\sim}{B}} = e^{\underset{\sim}{A}} e^{\underset{\sim}{B}} .$$

In connection with $\ln \underset{\sim}{A}$

$$\ln \underset{\sim}{A} = \frac{1}{2\pi i} \oint_{\Gamma} \frac{\ln z}{z-\underset{\sim}{A}} dz$$

we would like to demonstrate that

(3.2.28) $$e^{\ln \underset{\sim}{A}} = \underset{\sim}{A}.$$

We now demonstrate more generally that $f(g(z)) = F(z) \Rightarrow f(g(\underset{\sim}{A})) = F(\underset{\sim}{A})$, provided all quantities are properly defined, i.e., $g(z)$ is analytic in the neighborhood of the eigenvalues of $\underset{\sim}{A}$, and f in turn is analytic in the neighborhood of the eigenvalues of $g(\underset{\sim}{A})$. Therefore, we may first write

$$g(\underset{\sim}{A}) = \frac{1}{2\pi i} \oint_{\Gamma} \frac{g(z)dz}{z-\underset{\sim}{A}}$$

also

$$\frac{1}{\zeta-g(\underset{\sim}{A})} = \frac{1}{2\pi i} \oint_{\Gamma} \frac{1}{\zeta-g(z)} \frac{1}{(z-\underset{\sim}{A})} dz.$$

Multiplying by $\frac{f(\zeta)}{2\pi i}$ and integrating we obtain

$$f(g(\underset{\sim}{A})) = \frac{1}{2\pi i} \oint_{\gamma} \frac{f(\zeta)d\zeta}{\zeta-g(\underset{\sim}{A})}$$

$$= \frac{1}{2\pi i} \oint_{\gamma} f(\zeta)d\zeta \; \frac{1}{2\pi i} \oint_{\gamma} \frac{dz}{(\zeta-g(z))(z-\underset{\sim}{A})}$$

The curve (curves) enclose the eigenvalues of $g(\underset{\sim}{A})$. We choose Γ so that its image under $g(z)$ totally encloses γ, therefore interchanging order of integration

$$f(g(\underset{\sim}{A})) = \frac{1}{2\pi i} \oint_{\Gamma} \frac{dz}{z-A} \cdot \frac{1}{2\pi i} \oint_{\gamma} \frac{f(\zeta)}{\zeta-g(z)} d\zeta$$

(3.2.29)
$$= \frac{1}{2\pi i} \oint_{\Gamma} \frac{f(g(z))dz}{z-\underset{\sim}{A}}$$

$$= \frac{1}{2\pi i} \oint_{\Gamma} \frac{F(z)dz}{z-\underset{\sim}{A}}$$

$$= F(\underset{\sim}{A}).$$

This, in particular, demonstrates the validity of (3.2.28).

Before applying these developments to the solution of ordinary differential equations we make some remarks on the differentiation of matrices. As pointed out earlier

$$\left(\frac{d\underset{\sim}{A}}{dt}\right)_{ij} = \frac{d}{dt} (\underset{\sim}{A})_{ij}.$$

It is tempting to write

$$\frac{d}{dt}\Big[f(\underset{\sim}{A}(t))\Big] = f'(\underset{\sim}{A})\underset{\sim}{A}'(t)$$

however, this is incorrect in general. For example consider

$$\underset{\sim}{A}^n(t).$$

Differentiation gives

$$\frac{d}{dt}\underset{\sim}{A}^n(t) = \frac{d\underset{\sim}{A}}{dt}\cdot\underset{\sim}{A}^{n-1} + \underset{\sim}{A}\frac{d\underset{\sim}{A}}{dt}\underset{\sim}{A}^{n-2} + \cdots + \underset{\sim}{A}^{n-1}\frac{d\underset{\sim}{A}}{dt}\,.$$

Therefore, if $\underset{\sim}{A}'$ and $\underset{\sim}{A}$ commute the usual differentiation formula holds, and barring this there is no reason to assume its validity. We clearly have

$$\frac{d}{dx}e^{f(x)\underset{\sim}{A}} = f'(x)\underset{\sim}{A}e^{f(x)\underset{\sim}{A}} \tag{3.2.30}$$

$$\frac{d}{dx}x^{\underset{\sim}{A}} = \underset{\sim}{A}x^{\underset{\sim}{A}-\underset{\sim}{I}} \tag{3.2.31}$$

since the independent variable occurs in each through a scalar function.

3.3. Linear Ordinary Differential Equations with Constant Coefficients.

It follows from the previous section [3.2.20] that

$$\frac{d}{dz}e^{\underset{\sim}{A}z} = \underset{\sim}{A}e^{\underset{\sim}{A}z}$$

for a constant matrix. Hence the problem

$$(3.3.1) \qquad \begin{cases} \dfrac{d\underset{\sim}{w}}{dz} = \underset{\sim}{A}\underset{\sim}{w} \\ \underset{\sim}{w}(z_o) = \underset{\sim}{w}^o \end{cases}$$

for constant $\underset{\sim}{A}$ has the solution

$$(3.3.2) \qquad \underset{\sim}{w} = e^{\underset{\sim}{A}(z-z_o)}\underset{\sim}{w}^o.$$

Also from (3.1.14) the problem

$$(3.3.3) \qquad \begin{cases} \dfrac{d\underset{\sim}{w}}{dz} = \underset{\sim}{A}\underset{\sim}{w} + \underset{\sim}{f}(z) \\ \underset{\sim}{w}(z_o) = \underset{\sim}{w}^o \end{cases}$$

has the solution

$$(3.3.4) \qquad \underset{\sim}{w}(z) = e^{\underset{\sim}{A}(z-z_o)}\underset{\sim}{w}^o + \int_{z_o}^{z} e^{\underset{\sim}{A}(z-\zeta)}\underset{\sim}{f}(\zeta)d\zeta.$$

<u>Exercise 70.</u> Solve

(a) $$\frac{d}{dz}\begin{bmatrix} x \\ y \end{bmatrix} = \begin{bmatrix} a & b \\ c & d \end{bmatrix}\begin{bmatrix} x \\ y \end{bmatrix}$$

with

$$\begin{bmatrix} x \\ y \end{bmatrix}_{z=o} = \begin{bmatrix} \alpha \\ \beta \end{bmatrix}$$

(b) Suppose $a = \cos\theta$, $b = \sin\theta$

$c = -\sin\theta$, $d = \cos\theta$

in part (a).

Exercise 71. Solve

$$\frac{d}{dt}\begin{bmatrix} x \\ y \\ z \end{bmatrix} = \begin{bmatrix} 6 & 2 & -2 \\ -2 & 2 & 2 \\ 2 & 2 & 2 \end{bmatrix}\begin{bmatrix} x \\ y \\ z \end{bmatrix}$$

$$\begin{bmatrix} x \\ y \\ z \end{bmatrix}_{t=0} = \begin{bmatrix} x_o \\ y_o \\ z_o \end{bmatrix} .$$

Another type of problem which can be directly dealt with is

(3.3.5)
$$\frac{d}{dt}\underset{\sim}{X} = \underset{\sim}{A}\underset{\sim}{X} + \underset{\sim}{X}\underset{\sim}{B}$$
$$\underset{\sim}{X}\Big|_{t=0} = \underset{\sim}{X}_o$$

for the matrix $\underset{\sim}{X}$, with the matrices $\underset{\sim}{A}$ and $\underset{\sim}{B}$ constant. To solve, set

$$\underset{\sim}{X} = e^{\underset{\sim}{A}t}\underset{\sim}{Y}, \Rightarrow \underset{\sim}{Y}\Big|_{t=0} = \underset{\sim}{X}_o .$$

Substituting into (3.3.5)

$$\frac{d\underset{\sim}{Y}}{dt} = \underset{\sim}{Y}\underset{\sim}{B}$$

which has the solution

$$\underset{\sim}{Y} = \underset{\sim}{X}_o e^{\underset{\sim}{B}t}$$

and hence

(3.3.6)
$$\underset{\sim}{X} = e^{\underset{\sim}{A}t}\underset{\sim}{X}_o e^{\underset{\sim}{B}t} .$$

Exercise 72. Prove that a unique solution for $\underset{\sim}{X}$ of

$$\underset{\sim}{A}\underset{\sim}{X} + \underset{\sim}{X}\underset{\sim}{B} = \underset{\sim}{C}, \quad C \neq 0$$

is

$$\underset{\sim}{X} = -\int_0^\infty e^{\underset{\sim}{A}t} \underset{\sim}{C} e^{\underset{\sim}{B}t} dt$$

if this exists. (Hint: Use Equations 3.3.5 and 3.3.6)

To conclude this section we recall some methods for the solution of a single ODE with constant coefficients.

In an obvious way we associate with the polynomial

$$p(x) = x^n + a_1 x^{n-1} + \cdots + a_n$$

the differential operator (with constant coefficients)

$$p(\frac{d}{dz})$$

and we consider the differential equation

$$p(\frac{d}{dz})Y = 0. \tag{3.3.7}$$

Solutions are sought by supposing that $Y \sim e^{\lambda t}$, and

$$p(\frac{d}{dz})e^{\lambda z} = p(\lambda)e^{\lambda z} \tag{3.3.8}$$

and hence $e^{\lambda_i z}$ is solution if $p(\lambda_i) = 0$. If $\lambda_1, \ldots, \lambda_n$ represent n different roots of the nth order polynomial

$$p(\lambda) = 0 \tag{3.3.9}$$

then we can

Exercise 73. Prove $e^{\lambda_1 z},\ldots,e^{\lambda_n z}$ are n linearly independent solutions of (3.3.7).

Therefore, the general solution of (3.3.8) is given by

$$Y = \sum_{i=1}^{n} a_i e^{\lambda_i z}$$

where the a_i are to be determined by the initial data.

Next, suppose λ_0 is a root of multiplicity $m \leq n$ of (3.3.7). To treat this case we note from (3.3.8) that

$$\frac{\partial^k}{\partial\lambda^k} p(\tfrac{d}{dz})e^{\lambda z} = p(\tfrac{d}{dz})(z^k e^{\lambda z})$$
$$= \frac{\partial^k}{\partial\lambda^k}(p(\lambda)e^{\lambda z}).$$

Therefore,

$$z^k e^{\lambda_0 z}, \qquad k = 0,\ldots,m - 1,$$

are solutions to (3.3.7). In this way we can always generate n linearly independent solutions of (3.3.7).

Next, we consider the inhomogeneous problem,

$$p(\tfrac{d}{dz})Y = f(z). \tag{3.3.10}$$

In this connection it is useful to introduce the idea of an inverse operator, $(\frac{d}{dz} - \lambda)^{-1}$. This we do through the relationship

$$(\tfrac{d}{dz} - \lambda)(\tfrac{d}{dz} - \lambda)^{-1} = 1.$$

By direct differentiation one sees that

$$(\frac{d}{dz} - \lambda)^{-1} f = \int^z e^{\lambda(z-s)} f(s)ds$$

where the right hand side is an indefinite integral. To find a solution of (3.3.10) we first factor $p(\frac{d}{dz})$ and write

$$p(\frac{d}{dz})Y = \prod_{i=1}^{n} (\frac{d}{dz} - \lambda_i)Y = f(z)$$

where the λ_i are necessarily distinct. Then formally

$$Y_p = \prod_{i=n}^{1} (\frac{d}{dz} - \lambda_i)^{-1} f(z).$$

The subscript p is meant to indicate that this is a particular solution, that should be added to the general solution of (3.3.7). Applying the above considerations

$$\begin{aligned} Y_p &= \int^z e^{\lambda_n(z-z_n)} dz_n \cdots \int^{z_3} e^{\lambda_2(z_3-z_2)} dz_2 \int^{z_2} e^{\lambda_1(z_2-z_1)} f(z_1)dz_1 \\ &= e^{\lambda_n z} \int^z e^{-(\lambda_n-\lambda_{n-1})z_n} dz_n \cdots \int^{z_3} e^{-(\lambda_2-\lambda_1)z_2} dz_2 \int^{z_2} e^{-\lambda_1 z_1} f(z_1)dz_1. \end{aligned} \tag{3.3.11}$$

This for, (3.3.11), may always be reduced to a one dimensional integral, in fact,

Exercise 74. Denoting the roots of (3.3.8) and their multiplicities by $\lambda_1,\ldots,\lambda_k;\ \mu_1,\ldots,\mu_k\ (\sum_{i=1}^{k} \mu_i = n)$ reduce (3.3.11) to a one dimensional integral. (Take the lower limit of integration to be zero.)

Exercise 75. Show

$$(\frac{d}{dz} - \lambda_1)^{-1}(\frac{d}{dz} - \lambda_1) = (\frac{d}{dz} - \lambda_1)(\frac{d}{dz} - \lambda_1)^{-1}$$

$$(\frac{d}{dz} - \lambda_1)^{-1}(\frac{d}{dz} - \lambda_2)^{-1} = (\frac{d}{dz} - \lambda_2)^{-1}(\frac{d}{dz} - \lambda_1)^{-1}.$$

To conclude we briefly review some special techniques for certain forms of the inhomogeneous term $f(z)$ in (3.3.11).

Consider

$$p(\frac{d}{dz})Y = e^{\lambda_o z}$$

with $p(\lambda_o) \neq 0$, then clearly,

$$Y_p = \frac{e^{\lambda_o z}}{p(\lambda_o)} .$$

If λ_o is zero of multiplicity k of $p(\lambda)$, then differentiating (3.3.8) with respect to the parameter λ we obtain

$$\begin{aligned} \frac{\partial^k}{\partial\lambda^k} (p(\frac{d}{dz})e^{\lambda z}) &= \frac{\partial^k}{\partial\lambda^k} (p(\lambda)e^{\lambda z}) \\ &= p(d_z)(z^k e^{\lambda z}). \end{aligned}$$

Writing $p^{(k)}(\lambda_o) = \frac{\partial^k}{\partial\lambda^k} p(\lambda)\Big|_{\lambda=\lambda_o}$ we see that in this case

$$Y_p = \frac{z^k e^{\lambda_o z}}{p^{(k)}(\lambda_o)} .$$

Exercise 76. Find particular solutions of

$$(d_z^2-5d_z+6)y = \sin 3z + \cos 4z$$
$$(d_z^4-d_z^2)y = \sin z.$$

Another technique which is sometimes useful is obtained by first writing

$$\begin{aligned} p(d_z) &= p(d_z-a+a) \\ &= P(d_z-a), \end{aligned}$$

where P is also a polynomial of degree n. Then

$$\begin{aligned} p(d_z)e^{az}\phi(z) &= P(d_z-a)e^{az}\phi \\ &= e^{az}P(d_z)\phi \\ &= e^{az}p(d_z+a)\phi \end{aligned}$$

or formally

$$(p(d_z)e^{az}) = e^{az}p(d_z+a).$$

<u>Exercise 77</u>. Find particular solutions of

$$(d_z^2+5d_z+4)y = e^{-z}z^2$$

$$d_z^3y = e^z\ (\sin z + \cos z).$$

3.4. <u>Classification and General Properties of Ordinary Differential Equations in the Neighborhood of Singular Points.</u>

In the following two sections we consider

$$\frac{d\underset{\sim}{w}}{dz} = \underset{\sim}{B}(z)\underset{\sim}{w} \tag{3.4.1}$$

for $\underset{\sim}{B}(z)$ single-valued but singular. For convenience we take the singular point to be the origin (singular points at infinity are discussed later in this section). This of course involves no loss in generality. Our goal is to discuss the effect of such a singularity on solutions of (3.4.1).

The simplest representative of (3.4.1) having a singularity at the origin is

$$z^p \frac{d}{dz}\underset{\sim}{w} = \underset{\sim}{A}_0\underset{\sim}{w} \tag{3.4.2}$$

where $\underset{\sim}{A}_o$ is a constant matrix, and p is a positive integer. The solution of this equation follows directly from our considerations in section 3.2, in particular, (3.2.30, 31). The fundamental matrix is given by

$$\underset{\sim}{X} = z^{\underset{\sim}{A}_o}, \quad p = 1 \tag{3.4.3}$$

and

$$\begin{gathered} \underset{\sim}{X} = \exp[z^{1-p}\underset{\sim}{A}_o/(1-p)] \\ p = 2,3,\ldots \, . \end{gathered} \tag{3.4.4}$$

We now consider the various possibilities when the order, n, of $\underset{\sim}{A}_o$ is 2.

<u>Case 1.</u>

$$\underset{\sim}{A}_o = \underset{\sim}{S}\underset{\sim}{J}\underset{\sim}{S}^{-1}$$

$$\underset{\sim}{J} = \begin{bmatrix} \lambda & 1 \\ 0 & \lambda \end{bmatrix} = \lambda\underset{\sim}{I} + \underset{\sim}{K}, \quad \underset{\sim}{K} = \begin{bmatrix} 0 & 1 \\ 0 & 0 \end{bmatrix}$$

then if $p = 1$

$$\begin{aligned} z^{\underset{\sim}{A}_o} &= \underset{\sim}{S}z^{\underset{\sim}{J}}\underset{\sim}{S}^{-1} = \underset{\sim}{S}e^{(\ln z)\underset{\sim}{J}}\underset{\sim}{S}^{-1} \\ &= \underset{\sim}{S}e^{\lambda\underset{\sim}{I}(\ln z)} \cdot e^{\underset{\sim}{K}(\ln z)}\underset{\sim}{S}^{-1}. \end{aligned}$$

But $\underset{\sim}{K}$ is nilpotent $\underset{\sim}{K}^2 = 0$, and hence

$$\begin{aligned} z^{\underset{\sim}{A}_o} &= \underset{\sim}{S}z^{\lambda}\underset{\sim}{S}^{-1} + \underset{\sim}{S}z^{\lambda} \ln z \, \underset{\sim}{K}\underset{\sim}{S}^{-1} \\ z^{\underset{\sim}{A}_o} &= \underset{\sim}{S}z^{\lambda}\underset{\sim}{S}^{-1} + \underset{\sim}{S}z^{\lambda} \ln z \, \underset{\sim}{J}\underset{\sim}{S}^{-1} \\ &\quad - \underset{\sim}{S}z^{\lambda}\lambda\underset{\sim}{I} \ln z \, \underset{\sim}{S}^{-1} \\ z^{\underset{\sim}{A}_o} &= (z^{\lambda} - \lambda z^{\lambda} \ln z)\underset{\sim}{I} + z^{\lambda} \ln z \, \underset{\sim}{A}_o. \end{aligned} \tag{3.4.5}$$

If $p > 1$

$$\exp[z^{1-p}\underset{\sim}{A}_o/(1-p)] = \underset{\sim}{S}\exp[z^{1-p}\underset{\sim}{J}/(1-p)]\underset{\sim}{S}^{-1}$$

and in the same way as above

$$\exp[z^{1-p}\underset{\sim}{A}_o/(1-p)] = e^{\lambda z^{1-p}/(1-p)}\underset{\sim}{I}\left(1 - \frac{\lambda z^{1-p}}{1-p}\right) + e^{\lambda z^{1-p}/(1-p)}\frac{z^{1-p}}{1-p}\underset{\sim}{A}_o . \tag{3.4.6}$$

When $n > 2$, a representation in terms of $\underset{\sim}{I}$ and $\underset{\sim}{A}_o$ is not in general possible.

Case 2. $\underset{\sim}{A}_o = \underset{\sim}{S}\underset{\sim}{D}\underset{\sim}{S}^{-1}$ where $\underset{\sim}{D} = \begin{bmatrix} \lambda & 0 \\ 0 & \mu \end{bmatrix}$

but the eigenvalues λ, μ are not necessarily distinct.

Then if $p = 1$

$$z^{\underset{\sim}{A}_o} = \underset{\sim}{S}z^{\underset{\sim}{D}}\underset{\sim}{S}^{-1} = \underset{\sim}{S}\begin{bmatrix} z^{\lambda} & 0 \\ 0 & z^{\mu} \end{bmatrix}\underset{\sim}{S}^{-1} \tag{3.4.7}$$

and if $p > 1$

$$\exp[z^{1-p}\underset{\sim}{A}_o/(1-p)] = \underset{\sim}{S}\begin{bmatrix} \exp(\lambda z^{1-p}/(1-p)) & 0 \\ 0 & \exp(\mu z^{1-p}/(1-p) \end{bmatrix}\underset{\sim}{S}^{-1}. \tag{3.4.8}$$

When $n > 2$, the representation is clearly the same as (3.4.7) and (3.4.8).

Exercise 78. Find the general solution of

$$z^p \frac{d\underset{\sim}{w}}{dz} = \begin{bmatrix} -1 & 2 \\ -2 & 3 \end{bmatrix} \underset{\sim}{w}$$

for all integer $p \geq 1$.

Exercise 79. Find the general solution of

$$z^p \frac{d\underset{\sim}{w}}{dz} = \begin{bmatrix} \frac{1}{2} & 3 \\ \frac{1}{2} & 1 \end{bmatrix} \underset{\sim}{w}$$

for all inter $p \geq 1$.

Exercise 80. Represent the general solution of

$$z^p \frac{d\underset{\sim}{w}}{dz} = \underset{\sim}{A}_o \underset{\sim}{w}$$

for all integer $p \geq 1$, when order $\underset{\sim}{A}_o = 3$ and $\underset{\sim}{A}_o$ has at least two distinct eigenvalues.

Several points in the above discussion are worth noting. We see that when $p = 1$, the origin can become a branch point of the solutions. If the eigenvalues of $\underset{\sim}{A}_o$ are not distinct (3.4.5) shows the appearance of $\ln z$ (when $n > 2$, higher power of $\ln z$ can enter). Branch points can also be introduced by non-integer eigenvalues of $\underset{\sim}{A}_o$. This is demonstrated by (3.4.5) and (3.4.7). If the Jordan form is diagonal and if the eigenvalues are integers (positive or negative) the solution of (3.4.2) for $p = 1$ is single-valued with at most a pole at the origin.

The case $p \geq 2$ leads to solutions which are analytically very much different than for $p = 1$. Both (3.4.6) and (3.4.8) show that in this case the origin is an essential singularity. (Though the solutions are single-valued.) The same sort of behavior is encountered in the general case and in anticipation of this we make the following two definitions.

Definition. If the pole of $\underset{\sim}{B}(z)$ in (3.4.1) is order one, (3.4.1) is said to have a regular singular point at the origin.

Definition. If the pole of $\underset{\sim}{B}(z)$ in (3.4.1) is of order greater than one (3.4.1) is said to have an irregular singular point at the origin.

It is of interest to note that in the original German literature the terms "schwach singulare Stelle" and "stark singulare Stelle" are used. Their translations, i.e., weak singularity and strong singularity, would seem to more accurately describe the situation.

Note that if $\underset{\sim}{B}(z)$ is analytic at the origin, this is referred to as an analytic point. This is the case treated in section 3.0.

Circuit Relations.

In the event that the origin is a branch point of solutions of (3.4.1), a general characterization can be given. Consider the matrix form of (3.4.1)

$$\frac{d}{dz}\underset{\sim}{X} = \underset{\sim}{B}(z)\underset{\sim}{X} \tag{3.4.9}$$

where $\underset{\sim}{B}(z)$ is possibly singular at the origin - but single-valued there. Let us denote the operation of making a single circuit about the origin by the subscript Θ, thus $f(ze^{2\pi i}) = f_\Theta(z)$. By assumption

$$\underset{\sim}{B}_\Theta = \underset{\sim}{B}.$$

Performing this operation on (3.4.9)

$$\frac{d}{dz}\underset{\sim}{X}_\Theta = \underset{\sim}{B}(z)\underset{\sim}{X}_\Theta$$

so that $\underset{\sim}{X}_\Theta$ is a solution of (3.4.9). If $\underset{\sim}{X}$ is a fundamental solution, Theorem 313 implies that $\underset{\sim}{X}_\Theta$ is also a fundamental solution. Hence there exists a non-singular constant matrix C such that

$$\underset{\sim}{X}_{\circlearrowleft} = \underset{\sim}{X}\underset{\sim}{C}.$$

Next, suppose $\underset{\sim}{Y}(z)$ is also a fundamental solution of (3.4.9) then there exists a non-singular constant matrix $\underset{\sim}{T}$ such that

$$\underset{\sim}{Y} = \underset{\sim}{X}\underset{\sim}{T}$$

and performing a circuit

$$\underset{\sim}{Y}_{\circlearrowleft} = \underset{\sim}{X}_{\circlearrowleft}\underset{\sim}{T} = \underset{\sim}{X}\underset{\sim}{C}\underset{\sim}{T} = \underset{\sim}{Y}\underset{\sim}{T}^{-1}\underset{\sim}{C}\underset{\sim}{T}$$

and hence $\underset{\sim}{C}$ transforms under similarity.

Finally, let us define a constant matrix $\underset{\sim}{K}$ by

$$\underset{\sim}{K} = \frac{\ln \underset{\sim}{C}}{2\pi i}$$

then

$$\underset{\sim}{C} = e^{2\pi i \underset{\sim}{K}}$$

and consider the matrix function

$$\underset{\sim}{Q}(z) = \underset{\sim}{X}(z) z^{-\underset{\sim}{K}} = \underset{\sim}{X} e^{-\underset{\sim}{K} \ln z}$$

On performing a circuit about the origin

$$\begin{aligned} \underset{\sim}{Q}_{\circlearrowleft}(z) &= \underset{\sim}{X}_{\circlearrowleft} e^{-\underset{\sim}{K} \ln z - 2\pi i \underset{\sim}{K}} \\ &= \underset{\sim}{Q} \end{aligned}$$

and hence $\underset{\sim}{Q}$ is a single-valued at the origin. Hence we have the following representation of a fundamental matrix of (3.4.9),

$$\underset{\sim}{X} = \underset{\sim}{Q}(z)z^{\underset{\sim}{K}}$$

in terms of the single-valued matrix function $\underset{\sim}{Q}$ and the possibly many-valued $z^{-\underset{\sim}{K}}$. $\underset{\sim}{Q}$ of course may have an essential singularity at the origin.

In particular, the case (3.4.2) yields

$$\underset{\sim}{C} = e^{2\pi i \underset{\sim}{A}_o}, \qquad p = 1$$

$$\underset{\sim}{C} = \underset{\sim}{I}, \qquad p = 2,3,\dots .$$

Singular Points of an nth Order Scalar Ordinary Differential Equation.

We now classify the scalar ordinary differential equation

$$w^{(n)}(z) + p_1(z)w^{(n-1)}(z) + \cdots + p_n(z)w = 0. \tag{3.4.10}$$

Although our classifications will be made in accordance with the above definitions, we briefly consider their historical introduction.

Without loss of generality we consider the point of interest to be the origin. According to Fuch's definition, a regular solution is one that may be placed in the form

$$w = z^{\rho}[(\ln z)^r\phi_o(z) + (\ln z)^{r-1}\phi_1 + \cdots + \phi_r(z)] \tag{3.4.11}$$

where r is some integer, ρ is a constant, and each $\phi_i(z)$ analytic at the origin. He further defines (3.4.10) as having the origin as a regular singular point if all of its solutions are regular, i.e., of the form (3.4.11). Then we state without proof Fuch's Theorem. The origin, $z = 0$, is a regular singular point of (3.4.10) if and only if the coefficients $p_i(z)$ are such that

$$p_i(z) = \frac{q_i(z)}{z^i} \tag{3.4.12}$$

with $q_i(z)$ analytic at the origin.

To demonstrate the compatibility, with our previous definitions we write (3.4.10) as a system by defining

$$u_j = z^{j-1}w^{(j-1)}$$

so that

$$z\frac{du_j}{dz} = (j-1)z^{j-1}w^{(j-1)} + z^j w^{(j)}$$
$$= (j-1)u_j + u_{j+1}$$
$$j = 1,\ldots,n-1$$

and for $j = n$,

$$z\frac{du_n}{dz} = (n-1)u_n + z^n w^{(n)}$$
$$= (n-1)u_n - z^n\{\frac{q_1(z)}{z}w^{(n-1)} + \cdots + \frac{q_n}{z^n}w\}$$
$$= (n-1)u_n - q_1(z)u_n - q_2u_{n-1} - \cdots - q_nu_1.$$

Written in matrix form we have

$$z\frac{d}{dz}\begin{bmatrix} u_1 \\ u_2 \\ \vdots \\ \vdots \\ u_n \end{bmatrix} = \begin{bmatrix} 0 & 1 & 0 & 0 & 0 & \cdots & 0 \\ 0 & 1 & 1 & 0 & 0 & \cdots & 0 \\ 0 & 0 & 2 & 1 & 0 & \cdots & 0 \\ 0 & 0 & 0 & 3 & 1 & \cdots & 0 \\ & \cdots & \cdots & \cdots & \cdots & \cdots & \\ -q_n & -q_{n-1} & \cdots & \cdots & \cdots & \cdots & (n-1-q_1) \end{bmatrix}\begin{bmatrix} u_1 \\ u_2 \\ u_3 \\ u_4 \\ \vdots \\ u_n \end{bmatrix}$$

which by our previous definition has a regular singular point at the origin.

We, therefore, make the following:

<u>Definition</u>. Consider the equation (3.4.10) with single-valued coefficients $p_i(z)$, at least one of which is singular at the origin. Then if in (3.4.12) the $q_i(z)$

are analytic at the origin, equation (3.4.10) is said to have a regular singular point at the origin, otherwise the origin is said to be an irregular singular point of (3.4.10).

Solutions in the Neighborhood of Infinity.

We discuss (3.4.1) for $\underset{\sim}{B}(z)$ single-valued in the neighborhood of infinity. Setting

(3.4.13) $$z = \frac{1}{\zeta}$$

(3.4.1) becomes

$$\frac{d}{d\zeta}\underset{\sim}{w} = \frac{-1}{\zeta^2}\underset{\sim}{B}\left(\frac{1}{\zeta}\right)\underset{\sim}{w}.$$

Hence writing $\underset{\sim}{B}(z) = O(z^{p-2})$ for z large; ∞ is a regular point if $p \leq 0$; a regular singular point if $p = 1$; an irregular singular point if $p \geq 2$.

The same discussion can be applied to a single scalar equation. In particular, consider the second order equation

(3.4.14) $$\frac{d^2w}{dz^2} + p(z)\frac{dw}{dz} + q(z)w = 0.$$

Applying the transformation (3.4.13) to (3.4.14) we obtain

$$\frac{d^2w}{d\zeta^2} + \left\{\frac{2}{\zeta} - \frac{1}{\zeta^2}p\left(\frac{1}{\zeta}\right)\right\}\frac{dw}{d\zeta} + \frac{1}{\zeta^4}q\left(\frac{1}{\zeta}\right)w = 0.$$

Hence $z = \infty$ is a regular singular point if

$$p(z) = O\left(\frac{1}{z}\right)$$
$$q(z) = O\left(\frac{1}{z^2}\right)$$

as $|z| \to \infty$.

More generally one can

Exercise 81. Show that (3.4.10) has a regular singular point at infinity if each $p_i(z)$ is $O(z^{-i})$ at ∞.

The Equation.

$$z \frac{d\underset{\sim}{X}}{dz} = \underset{\sim}{A}_0 \underset{\sim}{X}. \tag{3.4.15}$$

As we pointed out earlier this equation has the fundamental matrix solution (3.4.3). It will be useful to further study this solution.

Consider the vector form of the equation

$$z \frac{d\underset{\sim}{w}}{dz} = \underset{\sim}{A}_0 \underset{\sim}{w}. \tag{3.4.16}$$

Then based on the form of $z^{\underset{\sim}{A}_0}$ we attempt a solution of (3.4.15) in the form

$$\underset{\sim}{w} = z^{\lambda} \underset{\sim}{w}_0$$

where $\underset{\sim}{w}_0$ is a constant vector. Substituting we find that

$$(\underset{\sim}{A}_0 - \lambda \underset{\sim}{I}) \underset{\sim}{w}_0 = 0$$

or that λ must be an eigenvalue of $\underset{\sim}{A}_0$ and $\underset{\sim}{w}_0$ its associated eigenvector. If by this process we generate n linearly independent eigenvectors, we obtain a fundamental matrix for (3.4.15). Alternately in this case $\underset{\sim}{A}_0$ is diagonalizable under similarity and the equation offers no real difficulty. It is rather the case when we do not generate n linearly independent eigenvectors that we want to further study. Clearly, in this case we have multiple eigenvalues.

Let us denote the eigenvalues of $\underset{\sim}{A}_0$ by $\lambda_1, \ldots, \lambda_k$, $k < n$ and, in particular, let the multiplicity of λ_1 be $m > 1$. Using the Dunford-Taylor representation we can write

$$z^{\underset{\sim}{A}_0} = \sum_{i=1}^{k} z_i^{\underset{\sim}{A}_o}$$

where

$$(3.4.17) \qquad z_i^{\underset{\sim}{A}_o} = \frac{1}{2\pi i} \oint_{\lambda_i} \frac{z^\zeta}{\zeta - \underset{\sim}{A}_o} d\zeta$$

In these the path of integration is a sufficiently small circle about the eigenvalue λ_i. Let us first note that each of the matrices $z_i^{\underset{\sim}{A}_o}$ is a solution of (3.4.15). To see this consider

$$\begin{aligned} z \frac{d}{dz} z_i^{\underset{\sim}{A}_o} &= \frac{1}{2\pi i} \oint_{\lambda_i} \frac{\zeta z^\zeta}{\zeta - \underset{\sim}{A}_o} d\zeta \\ &= \frac{1}{2\pi i} \oint_{\lambda_i} \frac{(\zeta - \underset{\sim}{A}_o) z^\zeta}{\zeta - A_o} d\zeta + \frac{1}{2\pi i} \oint_{\lambda_i} \frac{z^\zeta d\zeta}{\zeta - \underset{\sim}{A}_o} \underset{\sim}{A}_o \\ &= \underset{\sim}{A}_o z_i^{\underset{\sim}{A}_o}. \end{aligned}$$

In order to evaluate (3.4.17) say for $i = 1$ we first construct a polynomial $r(\zeta)$ of degree $m-1$ such that

$$\frac{d^k}{d\zeta^k} (z^\zeta - r(\zeta))\Big|_{\zeta=\lambda_1} = 0, \qquad k = 0,1,\ldots,m-1.$$

Writing

$$r(\zeta) = \alpha_o + \frac{\alpha_1(\zeta-\lambda_1)}{1!} + \cdots + \frac{\alpha_{m-1}(\zeta-\lambda_1)^{m-1}}{(m-1)!}$$

we find

$$\alpha_i = z^{\lambda_1}(\ln z)^i \qquad i = 0,1,\ldots,m-1.$$

Then from the same arguments leading to the interpolation formula (3.2.24) we find

$$z_1^{\underset{\sim}{A}_o} = z^{\lambda_1} \frac{1}{2\pi i} \left[\frac{d\zeta}{\zeta - \underset{\sim}{A}_o} + \frac{(\ln z)}{1!} \frac{1}{2\pi i} \oint_{\lambda_1} \frac{(\zeta-\lambda_1)}{\zeta - \underset{\sim}{A}_o} d\zeta \right.$$

$$+ \cdots + \frac{(\ln z)^{m-1}}{(m-1)!} \frac{1}{2\pi i} \oint_{\lambda_1} \frac{(\zeta-\lambda_1)^{m-1}}{\zeta-\underset{\sim}{A}_o} d\zeta \Bigg].$$

Next we note that

$$(\zeta-\lambda_1)^k = (\zeta-\underset{\sim}{A}_o+\underset{\sim}{A}_o-\lambda_1)^k = (\underset{\sim}{A}_o-\lambda_1)^k$$

$$+ (\zeta-\underset{\sim}{A}_o)[(k-1)(\underset{\sim}{A}_o-\lambda_1)^{k-1} + \cdots]$$

where the bracket represents a polynomial in $(\zeta-\underset{\sim}{A}_o)$. So that finally

$$z_1^{\underset{\sim}{A}_o} = z^{\lambda_1}\underset{\sim}{P}[1 + \frac{(\ln z)}{1!}(\underset{\sim}{A}_o-\lambda_1) + \frac{(\ln z)^2}{2!}(\underset{\sim}{A}_o-\lambda_1)^2 + \cdots + \frac{(\ln z)^{m-1}}{(m-1)!}(\underset{\sim}{A}_o-\lambda_1)^{m-1}] \tag{3.4.18}$$

with

$$\underset{\sim}{P} = \frac{1}{2\pi i}\oint_{\lambda_1} \frac{d\zeta}{\zeta-\underset{\sim}{A}_o}. \tag{3.4.19}$$

<u>Exercise 82.</u> Demonstrate that $\underset{\sim}{P}$ is projection operator

$$\underset{\sim}{P}^2 = \underset{\sim}{P}. \tag{3.4.20}$$

Let $\underset{\sim}{w}_1$ be such that

$$\underset{\sim}{A}_o\underset{\sim}{w}_1 = \lambda_1\underset{\sim}{w}_1.$$

Then from

$$(\underset{\sim}{A}_o-\zeta)\underset{\sim}{w}_1 = (\lambda_1-\zeta)\underset{\sim}{w}_1$$

we have

$$(\underset{\sim}{A}_o-\zeta)^{-1}\underset{\sim}{w}_1 = (\lambda_1-\zeta)^{-1}\underset{\sim}{w}_1.$$

Introducing this into the Dunford-Taylor integral demonstrates that $f(A_o)\underset{\sim}{w}_1 = f(\lambda_1)\underset{\sim}{w}_1$ and, in particular

$$\underset{\sim}{P}\underset{\sim}{w}_1 = \underset{\sim}{w}_1. \tag{3.4.21}$$

[Note if $\hat{\underset{\sim}{w}}$ is such that $\underset{\sim}{A}_o\hat{\underset{\sim}{w}} = \hat{\lambda}\hat{\underset{\sim}{w}}$, $\lambda_1 \neq \hat{\lambda}$ then the above discussions show that $\underset{\sim}{P}\underset{\sim}{w} = 0$.] Multiplying (3.4.18) on the right by $\underset{\sim}{w}_1$ gives the solution

$$z_1^{\underset{\sim}{A}_o}\underset{\sim}{w}_1 = z^{\lambda_1}\underset{\sim}{w}_1$$

which we already indicated above. However, from our solution in the form $z^{\underset{\sim}{A}_o}$ we know that there are m-independent solutions containing the factor z^{λ_1}. We now indicate how to find constant vectors $\underset{\sim}{w}_1^{(i)}$, $i = 1,\ldots,m$ such that $z_1^{\underset{\sim}{A}_o}\underset{\sim}{w}_1^{(i)}$, $i = 1,\ldots,m$ generate these m solutions.

Let $\underset{\sim}{w}_1^{(1)}$ represent an eigenvector corresponding to λ_1,

$$(\underset{\sim}{A}_o-\lambda_1)\underset{\sim}{w}_1^{(1)} = 0$$

and seek a solution to

$$(\underset{\sim}{A}_o-\lambda_1)\underset{\sim}{w}_1^{(2)} = \underset{\sim}{w}_1^{(1)}$$

such that $\underset{\sim}{P}\underset{\sim}{w}_1^{(2)} = \underset{\sim}{w}_1^{(2)}$. If this exists we next seek a solution to

$$(\underset{\sim}{A}_o-\lambda_1)\underset{\sim}{w}_1^{(3)} = \underset{\sim}{w}_1^{(2)}$$

such that $\underset{\sim}{P}\underset{\sim}{w}_1^{(3)} = \underset{\sim}{w}_1^{(3)}$ and so forth. It may then be proven that if this process is applied to each eigenvector of $\underset{\sim}{A}_o$, n linearly independent vectors are generated.

[Hence if a matrix has n linearly independent eigenvectors each such chain is of length one.] These are referred to as generalized eigenvectors.

Let the chain corresponding to $\underset{\sim}{w}_1^{(1)}$ be $\underset{\sim}{w}_1^{(i)}$, $i = 1,\ldots,m$. Then we may also characterize the chain by

$$(\underset{\sim}{A}_o - \lambda_1)^i \underset{\sim}{w}_1^{(i)} = 0$$

$$(\underset{\sim}{A}_o - \lambda_i)^{i-1} \underset{\sim}{w}_1^{(i)} \neq 0$$

Finally multiplying (3.4.18) on the right successively by $\underset{\sim}{w}_1^{(i)}$, $i = 1,\ldots,q$,

$$z_1^{\underset{\sim}{A}_o} w_1^{(1)} = z^{\lambda_1} \underset{\sim}{w}_1^{(1)}$$

$$z_1^{\underset{\sim}{A}_o} \underset{\sim}{w}_1^{(2)} = z^{\lambda_1} \{\underset{\sim}{w}_1^{(2)} + (\ln z)\underset{\sim}{w}_1^{(1)}\}$$

$$z_1^{\underset{\sim}{A}_o} \underset{\sim}{w}_1^{(3)} = z^{\lambda_1} \{\underset{\sim}{w}_1^{(3)} + (\ln z)\underset{\sim}{w}_1^{(2)} + \frac{(\ln z)^2}{2!} \underset{\sim}{w}_1^{(1)}\}$$

$$\cdots$$

$$z_1^{\underset{\sim}{A}_o} \underset{\sim}{w}_1^{(q)} = z^{\lambda_1} \{\underset{\sim}{w}_1^{(q)} + (\ln z)\underset{\sim}{w}_1^{(q-1)} + \frac{(\ln z)^2}{2!} \underset{\sim}{w}_1^{(q-2)} + \cdots + \frac{(\ln z)^{q-1} \underset{\sim}{w}_1^{(1)}}{(q-1)!}\}, \quad q < m.$$

Carrying through this procedure for each eigenvalue of $\underset{\sim}{A}_o$ we generate the n linearly independent solutions of (3.4.16).

<u>Exercise 83</u>. Find those solutions of

$$z \frac{d}{dz} \underset{\sim}{w} = \begin{bmatrix} -2 & 3 & -2 & 4 \\ 1 & 0 & 0 & 0 \\ 0 & 1 & 0 & 0 \\ 0 & 0 & 1 & 0 \end{bmatrix} \underset{\sim}{w}$$

which are bounded at the origin.

3.5. Linear Ordinary Differential Equations with Regular Singular Points.

We consider the equation

$$z \frac{d}{dz} \underset{\sim}{w} = \underset{\sim}{A}\underset{\sim}{w} \tag{3.5.1}$$

with $\underset{\sim}{A}(z)$ an $n \times n$ matrix analytic at the origin, i.e., it has the convergent power series expansion

$$\underset{\sim}{A}(z) = \sum_{i=o}^{\infty} \underset{\sim}{A}_i z^i < \infty, \quad |z| < |z_o| \quad \text{say} \tag{3.5.2}$$

We can directly seek a fundamental matrix of (3.5.1). Based on the considerations of section 3.4, we can look for a solution in the form

$$\underset{\sim}{X} = z^{\underset{\sim}{A}_o}(\underset{\sim}{P}_o + \underset{\sim}{P}_1 z + \underset{\sim}{P}_2 z^2 + \cdots) = z^{\underset{\sim}{A}_o}\underset{\sim}{P}(z).$$

If no two eigenvalues of $\underset{\sim}{A}_o$ differ by an integer, this leads to the determination of a fundamental matrix. In fact $\underset{\sim}{P}(z)$ is analytic - a fact already signaled by Fuch's theorem of last section. When two eigenvalues differ by an integer certain technical problems appear which may be overcome after the use of the so-called "shearing" transformations. [See [7] and [10] for this approach.] Although this is both concise and elegant we will instead seek vector solutions of (3.5.1). A major reason for this is that this is more useful in the applications where for example other considerations (e.g., boundedness at the origin) allows us to exclude from consideration all but a few of the vector solutions of (3.5.1).

Based on (3.4.3) we seek a solution of (3.5.1) in the form

$$\underset{\sim}{w} = z^{\lambda} \sum_{i=o}^{\infty} \underset{\sim}{w}_i(\lambda) z^i \tag{3.5.3}$$

where as indicated we allow the "constant" vectors to be functions of the parameter λ. Then formally

$$z \frac{d\underset{\sim}{w}}{dz} - \underset{\sim}{A}(z)\underset{\sim}{w} = \sum_{i=0}^{\infty} \{(\lambda+i)\underset{\sim}{w}_i - \sum_{\substack{m+n=i \\ m,n\geq 0}} \underset{\sim}{A}_m \underset{\sim}{w}_n\} z^{i+\lambda} .$$

We determine $\underset{\sim}{w}_k(\lambda)$ for $k \geq 1$ by

$$[\underset{\sim}{A}_0 - (\lambda+k)]\underset{\sim}{w}_k = -\sum_{m=1}^{k} \underset{\sim}{A}_m \underset{\sim}{w}_{k-m} \tag{3.5.4}$$

and therefore

$$z \frac{d\underset{\sim}{w}}{dz} - \underset{\sim}{A}\underset{\sim}{w} = z^{\lambda}(\lambda - \underset{\sim}{A}_0)\underset{\sim}{w}_o . \tag{3.5.5}$$

To make explicit the dependency on λ and $\underset{\sim}{w}_0$ we write

$$\underset{\sim}{w} = \underset{\sim}{w}(z;\ \lambda, \underset{\sim}{w}_o). \tag{3.5.6}$$

From (3.5.5) we see that (3.5.3) is a formal solution if $\lambda = \lambda_0$ say, is an eigenvalue and $\underset{\sim}{w}_0$ is the associated eigenvector. A difficulty appears in (3.5.4) however, if $\lambda_0 + k$ is also an eigenvalue for some integer k.

<u>Case 1.</u> No eigenvalues of $\underset{\sim}{A}_0$ differ by an integer.

To begin with if $\underset{\sim}{A}_0$ possesses n linearly independent eigenvectors $\underset{\sim}{w}_0^{(\mu)}$, $\mu = 1,\ldots,n$ then we generate n formal solutions

$$\underset{\sim}{w}^{(\mu)}(z;\ \lambda_\mu, \underset{\sim}{w}_0^{(\mu)}) = z^{\lambda_\mu} \sum_{i=0}^{\infty} \underset{\sim}{w}_i^{(\mu)} z^i = z^{\lambda_\mu} \underset{\sim}{w}(\mu)$$

where the $\underset{\sim}{w}_i^{(\mu)}$, $i \geq 1$ are determined through (3.5.4). Of course the λ_μ, $\mu = 1,\ldots,n$ are not necessarily distinct. Moreover, it follows from the Fuch's theorem that $\underset{\sim}{w}(\mu)$ is analytic at origin. (We are here making use of the equivalence of a scalar ordinary differential equation and a system as demonstrated in section 3.0.)

If $\underset{\sim}{A}_0$ does not possess n linearly independent eigenvectors then there exist multiple eigenvalues and generalized eigenvectors. To be specific let there correspond to λ_1 the following chain (see previous section)

$$(\underset{\sim}{A}_o-\lambda_1)\underset{\sim}{\omega}^{(1)} = 0$$

$$(\underset{\sim}{A}_o-\lambda_1)\underset{\sim}{\omega}^{(2)} = \underset{\sim}{\omega}^{(1)}$$

$$\vdots$$

$$(\underset{\sim}{A}_o-\lambda_1)\underset{\sim}{\omega}^{(m)} = \underset{\sim}{\omega}^{(m-1)}.$$

Next let us denote by $\underset{\sim}{\omega}^{(i)}$, $i = 1,\ldots,m$ the m formal expansions (3.5.3), gotten by taking $\underset{\sim}{w}_o^{(i)} = \underset{\sim}{\omega}^{(i)}$, i.e.,

(3.5.7) $$\underset{\sim}{w}^{(i)} = \underset{\sim}{w}^{i}(z;\ \lambda,\underset{\sim}{\omega}^{(i)}), \qquad i = 1,\ldots,m.$$

These choices make $\underset{\sim}{w}_o^{(i)}$ free of λ, i.e.,

$$\frac{\partial \underset{\sim}{w}_o^{(i)}}{\partial\lambda} = 0 \qquad i = 1,\ldots,m.$$

Differentiating (3.5.5) with respect to λ we find

(3.5.8)
$$\frac{\partial^p}{\partial\lambda^p}\left(z\frac{d}{dz}\underset{\sim}{w}-\underset{\sim}{A}\underset{\sim}{w}\right) = \left(z\frac{d}{dz}-\underset{\sim}{A}\right)\frac{\partial^p w}{\partial\lambda^p}$$
$$= \frac{\partial^p}{\partial\lambda^p}\left[z^\lambda(\lambda-\underset{\sim}{A}_o)\underset{\sim}{w}_o\right] = z^\lambda(\ln z)^p(\lambda-\underset{\sim}{A}_o)\underset{\sim}{w}_o + pz^\lambda(\ln z)^{p-1}\underset{\sim}{w}_o.$$

Finally, let $p < m$ and consider

$$\left(z\frac{d}{dz}-\underset{\sim}{A}\right)\left(\frac{1}{p!}\frac{\partial^p}{\partial\lambda^p}\underset{\sim}{w}^{(1)} + \frac{1}{(p-1)!}\frac{\partial^{p-1}}{\partial\lambda^{p-1}}\underset{\sim}{w}^{(2)} + \ldots + \underset{\sim}{w}^{(p+1)}\right)$$

$$= \frac{z^\lambda(\ln z)^p}{p!}(\lambda-\underset{\sim}{A}_o)\underset{\sim}{w}_o^{(1)} + z^\lambda(\ln z)^{p-1}\underset{\sim}{w}_o^{(1)}/(p-1)!$$

$$+ z^{\lambda}(\ln z)^{p-1}(\lambda - \underset{\sim}{A}_0) w_0^{(2)}/(p-1)!$$

$$+ z^{\lambda}(\ln z)^{p-2}\underset{\sim}{w}_0^{(2)}/(p-2)!$$

$$+ z^{\lambda}(\ln z)^{p-2}(\lambda - \underset{\sim}{A}_0)\underset{\sim}{w}_0^{(3)}/(p-2)!$$

$$+ z^{\lambda}(\ln z)^{p-3}\underset{\sim}{w}_0^{(3)}/(p-3)!$$

$$+ \cdot \cdot \cdot \cdot \cdot \cdot \cdot \cdot \cdot \cdot \cdot \cdot \cdot \cdot \cdot$$

$$+ z^{\lambda}(\ln z)(\lambda - \underset{\sim}{A}_0)\underset{\sim}{w}_0^{(p)}$$

$$+ z^{\lambda}\underset{\sim}{w}_0^{(p)} + z^{\lambda}(\lambda - \underset{\sim}{A}_0)\underset{\sim}{w}_0^{(p+1)}.$$

Hence, setting $\lambda = \lambda_1$ on the right hand side makes it vanish. Therefore, we have generated the m-independent solutions

$$(3.5.9)\begin{cases} \underset{\sim}{w}^{(1)}(z;\ \lambda_1) \\ \dfrac{\partial \underset{\sim}{w}^{(1)}}{\partial\lambda}(z;\ \lambda)\Big|_{\lambda=\lambda_1} + \underset{\sim}{w}^{(2)}(z;\ \lambda_1) \\ \vdots \\ \dfrac{1}{(m-1)!}\dfrac{\partial^{m-1}}{\partial\lambda^{m-1}}\underset{\sim}{w}^{(1)}(z;\ \lambda)\Big|_{\lambda=\lambda_1} + \dfrac{1}{(m-2)!}\dfrac{\partial^{m-2}}{\partial\lambda^{m-2}}\underset{\sim}{w}^{(2)}(z;\ \lambda)\Big|_{\lambda=\lambda_1} + \cdots + \underset{\sim}{w}^{(m)}(z;\ \lambda_1) \end{cases}$$

Carrying out the same procedure for each of the chains of λ_i, we generate n-independent solutions. Also as is implied by the Fuch's theorem, the resulting power series have a non-zero radius of convergence.

<u>Case 2.</u> There exist eigenvalues of $\underset{\sim}{A}_0$ differing by integers.

To be specific let us assume that $\lambda_0, \lambda_1, \ldots, \lambda_p$ are eigenvalues of $\underset{\sim}{A}_0$ arranged according to increasing value and such that $(\lambda_i - \lambda_0)$, $i = 1, \ldots, p$ are positive integers. We assume that there are no other eigenvalues having this property. Let the multiplicities of the λ_i be m_i,

$$m = \sum_{i=1}^{p} m_i.$$

Consider next the formal expansion gotten by taking

$$\underset{\sim}{w}_0 = (\lambda-\lambda_0)^m \underset{\sim}{\omega}_0$$

where $(A_0-\lambda_0)\underset{\sim}{\omega}_0 = 0$. Equation (3.5.5) now takes the form

$$(z \frac{d}{dz} - \underset{\sim}{A})\underset{\sim}{w} = z^{\lambda}(\lambda-\lambda_0)^m(\lambda-\underset{\sim}{A}_0)\underset{\sim}{\omega}_0.$$

Further consider equation (3.5.4) for the determination of the terms in the series of $\underset{\sim}{w}$.

$$[\underset{\sim}{A}_0 - (\lambda+k)]\underset{\sim}{w}_k = - \sum_{m=1}^{k} \underset{\sim}{A}_m \underset{\sim}{w}_{k-m}.$$

Since $\underset{\sim}{w}_0$ has the factor $(\lambda-\lambda_0)^m$ each $\underset{\sim}{w}_i$ will also have this factor — until k reaches $k = \lambda_1 - \lambda_0$. The matrix $[\underset{\sim}{A}_0 - (\lambda+\lambda_1-\lambda_0)]^{-1}$ has a singularity of order of m_1 at $\lambda = \lambda_0$. Hence $\underset{\sim}{w}_{(\lambda_1-\lambda_0)}$ will have a zero of order $(\lambda-\lambda_0)^{m_2+\cdots+m_p}$.

Proceeding in this way we determined all $\underset{\sim}{w}_i$ and these in fact exist at $\lambda = \lambda_0$. For by our choice of $\underset{\sim}{w}_0$, $\underset{\sim}{w}_{(\lambda_p-\lambda_0)}$ will exist and after this point the coefficient matrix $[A - (\lambda+k)]$ never again becomes singular at $\lambda = \lambda_0$.

The vector function $\underset{\sim}{w}(z; \lambda)$ determined in this way does indeed lead to solution when $\lambda = \lambda_0$. It is clear, however, that $\underset{\sim}{w}_i = 0$, $i = 1,\ldots,\lambda_p - \lambda_0 - 1$ when $\lambda = \lambda_0$, we in fact only determine the solution corresponding to $\lambda = \lambda_p$ in this way. To obtain a solution corresponding to $\lambda = \lambda_0$ we compute

$$\left.\frac{\partial^m}{\partial\lambda^m} \underset{\sim}{w}(z; \lambda)\right|_{\lambda=\lambda_0}$$

which is a solution and in fact the one we seek.

If λ_0 possesses generalized eigenvectors the same procedure must be pursued for each of these and a procedure similar to Case 1 adopted. We do not go into these details.

Example. Consider

$$z \frac{d}{dz} \underset{\sim}{w} = \begin{bmatrix} 1 & 1 & 0 \\ 0 & 1 & -z \\ -z & 0 & -1 \end{bmatrix} \underset{\sim}{w}$$

therefore,

$$\underset{\sim}{A}_0 = \begin{bmatrix} 1 & 1 & 0 \\ 0 & 1 & 0 \\ 0 & 0 & -1 \end{bmatrix}, \quad \underset{\sim}{A}_1 = \begin{bmatrix} 0 & 0 & 0 \\ 0 & 0 & -1 \\ -1 & 0 & 0 \end{bmatrix}, \quad \underset{\sim}{A}_k = 0, \quad k > 1.$$

The eigenvalues and eigenvectors of A_0 are

$$\lambda = -1, \qquad \underset{\sim}{w}_o = [0,0,1]$$

$$\lambda = 1, \qquad \underset{\sim}{w}_0 = [1,0,0], \text{ and the generalized eigenvector}$$

$$\tilde{\underset{\sim}{w}}_o = [0;1,0] \quad .$$

The solution is formally written as $\underset{\sim}{w} = z^{\lambda} \sum_{i=0}^{\infty} \underset{\sim}{w}_i(\lambda) z^i$ and due to form $\underset{\sim}{A}_i$ we have

$$(\underset{\sim}{A}_0 - \lambda - k)\underset{\sim}{w}_k = -\underset{\sim}{A}_1 \underset{\sim}{w}_{k-1}.$$

Solutions corresponding to $\lambda = 1$

$$\underset{\sim}{w}_0 = [1,0,0].$$

Next

$$(\underset{\sim}{A}_o-\lambda-1)\underset{\sim}{w}_1 = -\underset{\sim}{A}_1\underset{\sim}{w}_o \Rightarrow \begin{bmatrix} -\lambda & 1 & 0 \\ 0 & -\lambda & 0 \\ 0 & 0 & -2-\lambda \end{bmatrix}\begin{bmatrix} w_{11} \\ w_{12} \\ w_{13} \end{bmatrix} = \begin{bmatrix} 0 & 0 & 0 \\ 0 & 0 & 1 \\ 1 & 0 & 0 \end{bmatrix}\begin{bmatrix} 1 \\ 0 \\ 0 \end{bmatrix}$$

$$\underset{\sim}{w}_1 = [0,0,-\frac{1}{2+\lambda}]; \qquad \frac{\partial \underset{\sim}{w}_1}{\partial\lambda} = [0,0,\frac{1}{(2+\lambda)^2}].$$

And $\underset{\sim}{w}_2$ is determined from

$$\begin{bmatrix} -\lambda-1 & 1 & 0 \\ 0 & -\lambda-1 & 0 \\ 0 & 0 & -3-\lambda \end{bmatrix}\begin{bmatrix} w_{21} \\ w_{22} \\ w_{23} \end{bmatrix} = \begin{bmatrix} 0 & 0 & 0 \\ 0 & 0 & 1 \\ 1 & 0 & 0 \end{bmatrix}\begin{bmatrix} 0 \\ 0 \\ -\frac{1}{2+\lambda} \end{bmatrix}$$

$$\underset{\sim}{w}_2 = [\frac{1}{(\lambda+1)^2(\lambda+2)}, \frac{1}{(\lambda+1)(\lambda+2)}, 0];$$

$$\frac{\partial \underset{\sim}{w}_2}{\partial\lambda} = [-\frac{3\lambda+5}{(\lambda+1)^3(\lambda+2)^2}, -\frac{2\lambda+3}{(\lambda+1)^2(\lambda+2)^2}, 0]$$

and so forth.

Next compute the expansion based on the generalized eigenvector $\underset{\sim}{\tilde{w}}_o = [0,+1,0]$

$$(\underset{\sim}{A}_o-\lambda-1)\underset{\sim}{\tilde{w}}_1 = -\underset{\sim}{A}_1\underset{\sim}{\tilde{w}}_o \Rightarrow \begin{bmatrix} -\lambda & 1 & 0 \\ 0 & -\lambda & 0 \\ 0 & 0 & (-2-\lambda) \end{bmatrix}\begin{bmatrix} \tilde{w}_{11} \\ \tilde{w}_{12} \\ \tilde{w}_{13} \end{bmatrix} = \begin{bmatrix} 0 & 0 & 0 \\ 0 & 0 & 1 \\ 1 & 0 & 0 \end{bmatrix}\begin{bmatrix} 0 \\ +1 \\ 0 \end{bmatrix}$$

$$\underset{\sim}{\tilde{w}}_1 = 0$$

and hence $\underset{\sim}{\tilde{w}}_k = 0,\ k \geq 1$.

The two solutions corresponding to $\lambda = 1$ are

$$\underset{\sim}{w}(\lambda=1) = z\left\{\begin{bmatrix}1\\0\\0\end{bmatrix} + z\begin{bmatrix}0\\0\\-\frac{1}{3}\end{bmatrix} + z^2\begin{bmatrix}\frac{1}{12}\\\frac{1}{6}\\0\end{bmatrix} + O(z^3)\right\}$$

and

$$\left.\frac{\partial \underset{\sim}{w}}{\partial\lambda}\right|_{\lambda=1} + \tilde{\underset{\sim}{w}}(\lambda=1) = z \ln z\left\{\begin{bmatrix}1\\0\\0\end{bmatrix} + z\begin{bmatrix}0\\0\\-\frac{1}{3}\end{bmatrix} + z^2\begin{bmatrix}\frac{1}{12}\\\frac{1}{6}\\0\end{bmatrix} + \cdots\right\}$$

$$+\left\{z\begin{bmatrix}0\\0\\\frac{1}{9}\end{bmatrix} + z\begin{bmatrix}-\frac{1}{9}\\-\frac{5}{36}\\0\end{bmatrix} + \cdots\right\}$$

$$+ z\begin{bmatrix}0\\+1\\0\end{bmatrix}.$$

Solutions corresponding to $\lambda = -1$. Since this eigenvalue differs from the other by an integer and since $\lambda = 1$ is of n multiplicty, two, we take instead for $\underset{\sim}{w}_0$

$$\underset{\sim}{w}_0 = (\lambda+1)^2\begin{bmatrix}0\\0\\1\end{bmatrix}.$$

Then

$$(\underset{\sim}{A}_0-\lambda-1)\underset{\sim}{w}_1 = -\underset{\sim}{A}_1\underset{\sim}{w}_o$$

or

$$\begin{bmatrix} -\lambda & 1 & 0 \\ 0 & -\lambda & 0 \\ 0 & 0 & -2-\lambda \end{bmatrix} \begin{bmatrix} w_{11} \\ w_{12} \\ w_{13} \end{bmatrix} = (\lambda+1)^2 \begin{bmatrix} 0 & 0 & 0 \\ 0 & 0 & 1 \\ 1 & 0 & 0 \end{bmatrix} \begin{bmatrix} 0 \\ 0 \\ 1 \end{bmatrix}$$

$$= \begin{bmatrix} 0 \\ (\lambda+1)^2 \\ 0 \end{bmatrix}$$

and

$$\underset{\sim}{w}_1 = -[+\frac{(\lambda+1)^2}{\lambda^2}, \frac{(\lambda+1)^2}{\lambda}, 0].$$

Next

$$[\underset{\sim}{A}_0 - (\lambda+2)]\underset{\sim}{w}_2 = -\underset{\sim}{A}_1 w_1$$

$$\begin{bmatrix} -\lambda-1 & 1 & 0 \\ 0 & -\lambda-1 & 0 \\ 0 & 0 & -3-\lambda \end{bmatrix} \begin{bmatrix} w_{21} \\ w_{22} \\ w_{23} \end{bmatrix} = -\begin{bmatrix} 0 & 0 & 0 \\ 0 & 0 & 1 \\ 1 & 0 & 0 \end{bmatrix} \begin{bmatrix} +\frac{(\lambda+1)^2}{\lambda^2} \\ +(\lambda+1)^2/\lambda \\ 0 \end{bmatrix}$$

$$\underset{\sim}{w}_2 = [0, 0, \frac{(\lambda+1)^2}{(\lambda+3)\lambda^2}].$$

This shows that the factor $(\lambda+1)^2$ was unnecessary. Therefore, instead of carrying it and eventually differentiating the formal expression twice, we return to (3.5.3) and merely take

$$\underset{\sim}{w}_0 = \begin{bmatrix} 0 \\ 0 \\ 1 \end{bmatrix}$$

then

$$\underset{\sim}{w}_1 = [+\frac{1}{\lambda^2}, +\frac{1}{\lambda}, 0]$$

$$\underset{\sim}{w}_2 = [0, 0, \frac{1}{\lambda^2(\lambda+3)}]$$

and the corresponding solution is

$$\underset{\sim}{w} = z^{-1}\left\{ \begin{bmatrix} 0 \\ 0 \\ 1 \end{bmatrix} + z \begin{bmatrix} -1 \\ 1 \\ 0 \end{bmatrix} + z^2 \begin{bmatrix} 0 \\ 0 \\ \frac{1}{2} \end{bmatrix} + \cdots \right\}.$$

<u>Example</u>. Find the first few terms in the expansion of the solutions

$$z \frac{d\underset{\sim}{w}}{dz} = \begin{bmatrix} 1 & -z & -z \\ -z & 2 & -z \\ -z & -z & 3 \end{bmatrix} \underset{\sim}{w} .$$

<u>Solution</u>

$$\underset{\sim}{A}_0 = \begin{bmatrix} 1 & 0 & 0 \\ 0 & 2 & 0 \\ 0 & 0 & 3 \end{bmatrix} ; \qquad \underset{\sim}{A}_1 = \begin{bmatrix} 0 & -1 & -1 \\ -1 & 0 & -1 \\ -1 & -1 & 0 \end{bmatrix} ; \qquad \underset{\sim}{A}_k = 0,\ k \geq 2$$

Eigenvalues and eigenvectors of $\underset{\sim}{A}_o$

$$\lambda = 1, \quad [1,0,0]$$

$$\lambda = 2, \quad [0,1,0]$$

$$\lambda = 3, \quad [0,0,1]$$

(1) Solution corresponding to $\lambda = 3$.

Substitute $\lambda = 3$ directly since $\lambda = 3$ is simple and does not differ from another eigenvalue by an integer.

$$(\underset{\sim}{A}_0 - 3 - k)\underset{\sim}{w}_k = -\underset{\sim}{A}_1 \underset{\sim}{w}_{k-1}$$

$$\begin{bmatrix} -3 & 0 & 0 \\ 0 & -2 & 0 \\ 0 & 0 & -1 \end{bmatrix} \begin{bmatrix} w_{11} \\ w_{12} \\ w_{13} \end{bmatrix} = \begin{bmatrix} 0 & 1 & 1 \\ 1 & 0 & 1 \\ 1 & 1 & 0 \end{bmatrix} \begin{bmatrix} 0 \\ 0 \\ 1 \end{bmatrix}$$

$$\underset{\sim}{w}_1 = [-\frac{1}{3}, -\frac{1}{2}, 0]$$

$$\begin{bmatrix} -4 & 0 & 0 \\ 0 & -3 & 0 \\ 0 & 0 & -2 \end{bmatrix} \begin{bmatrix} w_{21} \\ w_{22} \\ w_{23} \end{bmatrix} = \begin{bmatrix} 0 & 1 & 1 \\ 1 & 0 & 1 \\ 1 & 1 & 0 \end{bmatrix} \begin{bmatrix} -\frac{1}{3} \\ -\frac{1}{2} \\ 0 \end{bmatrix}$$

$$\underset{\sim}{w}_2 = [\frac{1}{8}, \frac{1}{9}, \frac{5}{12}] \quad \text{etc.}$$

and the solution is

$$z^3 \left(\begin{bmatrix} 0 \\ 0 \\ 1 \end{bmatrix} + z \begin{bmatrix} -\frac{1}{3} \\ -\frac{1}{2} \\ 0 \end{bmatrix} + z^2 \begin{bmatrix} \frac{1}{8} \\ \frac{1}{9} \\ \frac{5}{12} \end{bmatrix} + \cdots \right)$$

(2) Solution corresponding to $\lambda = 2$.

The preceding eigenvalue from this by an integer and we, therefore, write

$$\underset{\sim}{w}_0 = (\lambda-2)\begin{bmatrix} 0 \\ 1 \\ 0 \end{bmatrix}$$

$$(\underset{\sim}{A}_0-\lambda-k)\underset{\sim}{w}_k = -\underset{\sim}{A}_1\underset{\sim}{w}_{k-1}$$

$k = 1$

$$\begin{bmatrix} -\lambda & 0 & 0 \\ 0 & 1-\lambda & 0 \\ 0 & 0 & 2-\lambda \end{bmatrix}\begin{bmatrix} w_{11} \\ w_{12} \\ w_{13} \end{bmatrix} = \begin{bmatrix} 0 & 1 & 1 \\ 1 & 0 & 1 \\ 1 & 1 & 0 \end{bmatrix}\begin{bmatrix} 0 \\ \lambda-2 \\ 0 \end{bmatrix}$$

$$\underset{\sim}{w}_1 = \begin{bmatrix} (2-\lambda)/\lambda \\ 0 \\ -1 \end{bmatrix}$$

$k = 2$

$$\begin{bmatrix} -\lambda-1 & 0 & 0 \\ 0 & -\lambda & 0 \\ 0 & 0 & 1-\lambda \end{bmatrix}\begin{bmatrix} w_{21} \\ w_{22} \\ w_{23} \end{bmatrix} = \begin{bmatrix} 0 & 1 & 1 \\ 1 & 0 & 1 \\ 1 & 1 & 0 \end{bmatrix}\begin{bmatrix} \frac{2-\lambda}{\lambda} \\ 0 \\ -1 \end{bmatrix}$$

$$\underset{\sim}{w}_2 = [\frac{1}{\lambda+1}, \frac{1}{\lambda}(1 - \frac{2-\lambda}{\lambda}), \frac{2-\lambda}{(1-\lambda)\lambda}], \text{ etc.}$$

The solution is then given by

$$\frac{\partial}{\partial\lambda}[z^{\lambda}\{(\lambda-2)\begin{bmatrix} 0 \\ 1 \\ 0 \end{bmatrix} + z\begin{bmatrix} (2-\lambda)/\lambda \\ 0 \\ -1 \end{bmatrix} +$$

$$+ z^2 \begin{bmatrix} \frac{1}{\lambda+1} \\ \frac{2(\lambda-1)}{\lambda^2} \\ \frac{2-\lambda}{\lambda(1-\lambda)} \end{bmatrix} + \cdots \}] \Bigg|_{\lambda=2}$$

$$= z^2 \ln z \left\{ z \begin{bmatrix} 0 \\ 0 \\ -1 \end{bmatrix} + z^2 \begin{bmatrix} \frac{1}{3} \\ \frac{1}{2} \\ 0 \end{bmatrix} + \cdots \right\}$$

$$+ z^2 \left\{ \begin{bmatrix} 0 \\ 1 \\ 0 \end{bmatrix} + z \begin{bmatrix} -\frac{1}{2} \\ 0 \\ 0 \end{bmatrix} , + z^2 \begin{bmatrix} -\frac{1}{9} \\ 0 \\ \frac{1}{2} \end{bmatrix} + \cdots \right\}$$

(3) Solution corresponding to $\lambda = 1$.

$$\underset{\sim}{w}_0 = (\lambda-1)^2 \begin{bmatrix} 1 \\ 0 \\ 0 \end{bmatrix}$$

$$(\underset{\sim}{A}_0 - \lambda - k)\underset{\sim}{w}_k = -\underset{\sim}{A}_1 \underset{\sim}{w}_{k-1}$$

$$\begin{bmatrix} -\lambda & 0 & 0 \\ 0 & 1-\lambda & 0 \\ 0 & 0 & 2-\lambda \end{bmatrix} \begin{bmatrix} w_{11} \\ w_{12} \\ w_{13} \end{bmatrix} = \begin{bmatrix} 0 & 1 & 1 \\ 1 & 0 & 1 \\ 1 & 1 & 0 \end{bmatrix} \begin{bmatrix} (\lambda-1)^2 \\ 0 \\ 0 \end{bmatrix}$$

$$\underset{\sim}{w}_1 = [0, (1-\lambda), \frac{(\lambda-1)^2}{2-\lambda}]$$

$$\begin{bmatrix} -\lambda-1 & 0 & 0 \\ 0 & -\lambda & 0 \\ 0 & 0 & 1-\lambda \end{bmatrix} \begin{bmatrix} w_{21} \\ w_{22} \\ w_{23} \end{bmatrix} = \begin{bmatrix} 0 & 1 & 1 \\ 1 & 0 & 1 \\ 1 & 1 & 0 \end{bmatrix} \begin{bmatrix} 0 \\ 1-\lambda \\ \frac{(\lambda-1)^2}{2-\lambda} \end{bmatrix}$$

$$\underset{\sim}{w}_2 = [\frac{\lambda-1}{\lambda+1} + \frac{(\lambda-1)^2}{(\lambda+1)(\lambda-2)}, \frac{(\lambda-1)^2}{(\lambda-2)\lambda}, 1].$$

The solution is then given by

$$\frac{\partial^2}{\partial\lambda^2}\left[z^\lambda(\lambda-1)^2 \begin{bmatrix} 1 \\ 0 \\ 0 \end{bmatrix} + z^{\lambda+1} \begin{bmatrix} 0 \\ (1-\lambda) \\ \frac{(\lambda-1)^2}{(2-\lambda)} \end{bmatrix} + z^{\lambda+2} \begin{bmatrix} \frac{\lambda-1}{\lambda+1} + \frac{(\lambda-1)^2}{(\lambda+1)(\lambda-2)} \\ \frac{(\lambda-1)^2}{\lambda(\lambda-2)} \\ 1 \end{bmatrix} + \cdots \right]\Bigg|_{\lambda=1}$$

$$= \frac{\partial}{\partial\lambda}\left\{ z^\lambda \ln z(\lambda-1)^2 \begin{bmatrix} 1 \\ 0 \\ 0 \end{bmatrix} + z^{\lambda+1}\ln z \begin{bmatrix} 0 \\ (1-\lambda) \\ \frac{(\lambda-1)^2}{2-\lambda} \end{bmatrix} \right.$$

$$\left. + z^{\lambda+2}\ln z \begin{bmatrix} \frac{\lambda-1}{\lambda+1} + \frac{(\lambda-1)^2}{(\lambda+1)(\lambda-2)} \\ \frac{(\lambda-1)^2}{(\lambda-2)\lambda} \\ 1 \end{bmatrix} + \cdots \right\}\Bigg|_{\lambda=1}$$

$$+ \frac{\partial}{\partial\lambda}\left\{ 2z^\lambda(\lambda-1) \begin{bmatrix} 1 \\ 0 \\ 0 \end{bmatrix} + z^{\lambda+1} \begin{bmatrix} 0 \\ -1 \\ \frac{2(\lambda-1)}{2-\lambda} + \frac{(\lambda-1)^2}{(2-\lambda)^2} \end{bmatrix} \right.$$

$$
+ z^{\lambda+2}\begin{bmatrix} \frac{1}{\lambda+1} - \frac{(\lambda-1)}{(\lambda+1)^2} + \frac{2(\lambda-1)}{(\lambda+1)(\lambda-2)} - \frac{(\lambda-1)^2\{2\lambda-1\}}{(\lambda+1)^2(\lambda-2)^2} \\ \frac{2(\lambda-1)}{\lambda(\lambda-2)} - \frac{2(\lambda-1)^3}{(\lambda-2)^2\lambda^2} \\ 0 \end{bmatrix}\Bigg\}_{\lambda=1}
$$

$$
= z^2\ln z\begin{bmatrix} 0 \\ -1 \\ 0 \end{bmatrix} + z^3(\ln z)^2\begin{bmatrix} 0 \\ 0 \\ 1 \end{bmatrix} + z^3(\ln z)\begin{bmatrix} \frac{1}{2} \\ 0 \\ 0 \end{bmatrix}
$$

$$
+ 2z\begin{bmatrix} 1 \\ 0 \\ 0 \end{bmatrix} + z^2\ln z\begin{bmatrix} 0 \\ -1 \\ 0 \end{bmatrix} + z^2\begin{bmatrix} 0 \\ 0 \\ 2 \end{bmatrix} + z^3(\ln z)\begin{bmatrix} \frac{1}{2} \\ 0 \\ 0 \end{bmatrix}
$$

$$
+ z^3\begin{bmatrix} -\frac{1}{4} \quad -\frac{1}{4} \quad -\frac{2}{2} \\ -2 \\ 0 \end{bmatrix} + \cdots
$$

$$
= 2z\begin{bmatrix} 1 \\ 0 \\ 0 \end{bmatrix} + z^2\ln z\begin{bmatrix} 0 \\ -2 \\ 0 \end{bmatrix} + z^2\begin{bmatrix} 0 \\ 0 \\ 2 \end{bmatrix} + z^3(\ln z)^2\begin{bmatrix} 0 \\ 0 \\ 1 \end{bmatrix}
$$

$$
+ z^3(\ln z)\begin{bmatrix} 1 \\ 0 \\ 0 \end{bmatrix} + z^3\begin{bmatrix} -\frac{3}{2} \\ -2 \\ 0 \end{bmatrix} + \cdots .
$$

The Case of a Scalar Ordinary Differential Equation with a Regular Singular Point.

For simplicity we consider the second order equation

$$\frac{d^2w}{dz^2} + p(z)\frac{dw}{dz} + q(z)w = 0 \tag{3.5.10}$$

with

$$zp(z) = \sum_{i=o}^{\infty} p_i z^i \tag{3.5.11}$$

$$z^2q(z) = \sum_{i=o}^{\infty} q_i z^i, \tag{3.5.12}$$

convergent in some circle enclosing the origin. The extension of the method to be presented to higher order equations is direct.

Method of Fröbenius.

The method given below goes by this name. Actually we have used the ideas of this method in the discussion of the system (3.5.1).

We attempt a solution in the form

$$w = z^\lambda \sum_{i=o}^{\infty} c_i z^i \tag{3.5.13}$$

which from Fuch's theorem we know to be convergent. Computing derivatives,

$$w' = \lambda z^{\lambda-1} \sum_{k=0}^{\infty} c_k z^k + z^{\lambda-1} \sum_{k=1}^{\infty} kc_k z^k$$

$$w'' = \lambda(\lambda-1)z^{\lambda-2} \sum_{k=0}^{\infty} c_k z^k + 2\lambda z^{\lambda-2} \sum_{k=1}^{\infty} kc_k z^k + z^{\lambda-2} \sum_{k=2}^{\infty} k(k-1)c_k z^k.$$

Substituting these as well as (3.5.11), (3.5.12) and (3.5.13) into our equation (3.5.10), we find

$$\frac{d^2w}{dz^2} + p(z)\frac{dw}{dz} + q(z)w = c_o z^{\lambda-2}(\lambda-\lambda_1)(\lambda-\lambda_2) \tag{3.5.14}$$

where we have written

$$\lambda(\lambda-1) + \lambda p_0 + q_0 = (\lambda-\lambda_1)(\lambda-\lambda_2)$$

and hence

$$q_0 = \lambda_1\lambda_2$$

$$p_0 = 1 - \lambda_1 - \lambda_2.$$

Also we have taken all coefficients of $z^{\lambda-k}$, $k \geq 1$ to vanish, i.e.

$$c_n\{(\lambda+n)(\lambda+n-1) + (\lambda+n)p_0 + q_0\} = -\sum_{s=0}^{n-1} c_s\{(\lambda+s)p_{n-s} + q_{n-s}\}.$$

On making use of the above relations amongst $q_0, p_0, \lambda_1\lambda_2$. This can also be written as

$$\{(\lambda+n-\lambda_1)(\lambda+n-\lambda_2)\}c_n = -\sum_{s=0}^{n-1} c_s\{(\lambda+s)p_{n-s} + q_{n-s}\}. \tag{3.5.15}$$

If the indicial equation $(\lambda-\lambda_1)(\lambda-\lambda_2) = 0$ has distinct roots $\lambda_1 \neq \lambda_2$ and not differing by an integer

$$w(z,\lambda = \lambda_1), \quad w(z,\lambda = \lambda_2)$$

are two independent solutions.

If $\lambda_1 = \lambda_2$, then

$$w(z,\lambda = \lambda_1), \quad \frac{\partial w}{\partial \lambda}(z;\lambda)\Big|_{\lambda=\lambda_1}$$

are independent solutions.

If $\lambda_2 - \lambda_1 = k$ a positive integer take

$$c_o = (\lambda-\lambda_1)\tilde{c}_o$$

and

$$w(z,\lambda = \lambda_1); \quad \frac{\partial w}{\partial\lambda}(z,\lambda)\Big|_{\lambda=\lambda_1}$$

are independent solutions.

With appropriate modification the same discussion can be applied to higher order equations.

Example. The hypergeometric differential equation is

$$z(1-z)\frac{d^2y}{dz^2} + \{\gamma - (\alpha+\beta+1)z\}\frac{dy}{dz} - \alpha\beta y = 0 \tag{3.5.16}$$

where α,β,γ are constants.

Before seeking solutions to this we motivate its form. Suppose we seek to find the most general second order ordinary differential equation analytic except at most at 3 points of the complex plane. We allow these to be regular singular points. Using a linear fractional map we may always transform the location of the three regular singular points to $(0,1,\infty)$. Therefore, the ordinary differential equation we are seeking has the form

$$\frac{d^2w}{dz^2} + \frac{r_o+r_1z}{z(z-1)}\frac{dw}{dz} + \frac{s_o+s_1z+s_2z^2}{z^2(z-1)^2}w = 0. \tag{3.5.17}$$

Next suppose one of the roots of the indicial equation zt $z = 0$ is a_1 and at $z = 1$ is a_2, then by setting $w = z^{a_1}(z-1)^{a_2}\tilde{w}$ we get an equation for $\tilde{w}$ which will have a zero root of the indicial equation at $z = 0$ and at $z = 1$. As is easily seen this is equivalent to taking $s_o = 0$ and $s_2 = -s_1$ in (3.5.17). Therefore, we may just consider

$$\frac{d^2w}{dz^2} + \frac{r_o+r_1z}{z(z-1)}\frac{dw}{dz} + \frac{s_1}{z^2(1-z)}\,w = 0.$$

Finally, there is the convention of representing r_o, r_1, s_1 in terms of the roots of the indicial equation at ∞, α, β and the remaining root at the origin $(1-\gamma)$. If this is done, this results in (3.5.16). A simple calculation then shows that the remaining root at $z = 1$ is $\gamma - \alpha - \beta$.

Equations having at most regular singular points are said to be of Fuch's type. For such equations of second order there is a useful notation due to Riemann, which we illustrate for the hypergeometric equation,

$$y(z) = P\begin{pmatrix} 0 & 1 & \infty & \\ 0 & 0 & \alpha & z \\ (1-\gamma) & (\gamma-\alpha-\beta) & \beta & \end{pmatrix}.$$

For the sake of simplicity assume that none of the expressions γ, $\gamma - \alpha - \beta$, $\alpha - \beta$ are integers. Then for example at the origin one solution of (3.5.16) is

$$\begin{aligned} y(z) &= F(\alpha,\beta,\gamma;\ z) \\ &= 1 + \sum_{n=1}^{\infty} a_n z^n. \end{aligned}$$

The determination of the coefficients a_n follows from (3.5.15) and

$$a_n = \frac{\Gamma(\alpha+n)\Gamma(\beta+n)\Gamma(\gamma)}{n!\Gamma(\alpha)\Gamma(\beta)\Gamma(\gamma+n)}$$

$$a_1 = \frac{\alpha\beta\gamma}{\gamma}, \quad a_2 = \frac{\alpha(\alpha+1)\beta(\beta+1)}{1\cdot 2\gamma(\gamma+1)}$$

$$\cdots\cdots$$

In this notation the second solution is found to be

$$y(z) = z^{1-\gamma}F(\alpha-\gamma+1,\beta-\gamma+1,2-\gamma;\ z).$$

Finally, we mention that the following identifications can be made

$$(1-z)^n = F(-n,1,1,z)$$

$$\ln(1-z) = zF(1,1,2,z)$$

$$\sin^{-1}z = zF(\tfrac{1}{2}, \tfrac{1}{2}, \tfrac{3}{2};\ z^2).$$

Also, many of the so-called special functions may be represented in this way, see [14], [15].

3.6. Irregular Singular Points.

The case of an irregular singular point is technically more complex than the case of a regular singular point - and in many respects it is more subtle. We adhere to the customary practice of considering the irregular singular point to be located at infinity. Thus according to the discussion on page 234 if in

$$\frac{d\underset{\sim}{w}}{dz} = \underset{\sim}{B}(z)\underset{\sim}{w} \tag{3.6.1}$$

$\underset{\sim}{B}$ is non-vanishing as $z \to \infty$, then infinity is an irregular singular point of the equation. Thus for $\underset{\sim}{A}_0$ a constant matrix,

$$\frac{d\underset{\sim}{X}}{dt} = z^r\underset{\sim}{A}_0\underset{\sim}{X} \Rightarrow \underset{\sim}{X} = \exp\left[\frac{z^r}{r+1}\underset{\sim}{A}_0\right]$$

or if $\underset{\sim}{B}$ is a one by one matrix, i.e., the scalar case, a solution is

$$X = \exp\left[\int^z B(z')dz'\right]$$

so that if

$$B = B_r z^r + B_{r-1} z^{n-1} + \cdots + B_o + B_{-1} z^{-1} + \cdots$$

$$X = \exp\left[\frac{z^{r+1}}{r+1} B_r + \cdots + B_o z + B_{-1} \ln z + \cdots\right].$$

From these examples of an equation having infinity as an irregular singular point, it is seen that the point at infinity becomes an essential singularity of the solution.

For the reasons given in section 3.5, we seek vector instead of matrix solutions of (3.6.1). [See [7] or [10] for the matrix treatment.] We write

$$\frac{d\underset{\sim}{w}}{dz} = z^r \underset{\sim}{A}(z)\underset{\sim}{w} \tag{3.6.2}$$

with $\underset{\sim}{A}$ bounded at infinity, and the integer $r \geq 0$. We also assume

$$\underset{\sim}{A}(z) \sim \underset{\sim}{A}_o + \frac{\underset{\sim}{A}_1}{z} + \frac{\underset{\sim}{A}_2}{z^2} + \cdots \tag{3.6.3}$$

where the $\underset{\sim}{A}_i$ are constant $n \times n$ matrices. Motivated by the examples above we attempt a formal solution of the type

$$\underset{\sim}{w} = e^{q(z)}[\underset{\sim}{u}_o + \frac{\underset{\sim}{u}_1}{z} + \frac{\underset{\sim}{u}_2}{z^2} + \cdots]$$

with

$$q(z) = \lambda_o \frac{z^{r+1}}{r+1} + \lambda_1 \frac{z^r}{r} + \cdots + \lambda_r z + \lambda_{r+1} \ln z$$

and $\underset{\sim}{u}_i$ constant vectors. Substituting into (3.6.2) we formally obtain

$$(\lambda_o z^r + \lambda_1 z^{r-1} + \cdots + \lambda_r + \frac{\lambda_{r+1}}{z})(\underset{\sim}{u}_o + \frac{\underset{\sim}{u}_1}{z} + \cdots) - (\frac{\underset{\sim}{u}_1}{z^2} + \frac{2\underset{\sim}{u}_2}{z^3} + \cdots)$$

$$= z^r(\underset{\sim}{A}_o + \frac{\underset{\sim}{A}_1}{z} + \cdots)(\underset{\sim}{u}_o + \frac{\underset{\sim}{u}_1}{z} + \cdots).$$

Collecting terms and defining $\lambda_k = 0,\ k > r+1$, we have,

$$z^r \sum_{k=0}^{\infty} z^{-k} \sum_{i+j=k} (\lambda_i - \underset{\sim}{A}_i)\underset{\sim}{u}_j = \sum_{k=1}^{\infty} kz^{-k-1}\underset{\sim}{u}_k .$$

From this we obtain

$$(3.6.4) \qquad (\underset{\sim}{A}_o - \lambda_o)\underset{\sim}{u}_k = \sum_{i=1}^{k} (\lambda_i - \underset{\sim}{A}_i)\underset{\sim}{u}_{k-i} \qquad k = 0,\ldots,r+1$$

and

$$(3.6.5) \qquad (\underset{\sim}{A}_o - \lambda_o)\underset{\sim}{u}_k = (r+1-k)\underset{\sim}{u}_{k-r-1} + \sum_{i=1}^{k} (\lambda_i - \underset{\sim}{A}_i)\underset{\sim}{u}_{k-i} \qquad k = r+2,\ldots .$$

In order to determine the formal solution we will first suppose that $\underset{\sim}{A}_o$ has distinct eigenvalues. Let μ_o represent one such eigenvalue and also $\underset{\sim}{\omega}_o, \underset{\sim}{\tilde{\omega}}_o$ be the corresponding right and left eigenvectors, i.e.,

$$(3.6.6) \qquad (\underset{\sim}{A}_o - \mu_o)\underset{\sim}{\omega}_o = 0 = (\underset{\sim}{A}_o - \mu_o)^{+}\underset{\sim}{\tilde{\omega}}_o = 0.$$

Then taking

$$(3.6.7) \qquad \lambda_o = \mu_o, \quad \underset{\sim}{u}_o = \underset{\sim}{\omega}_o$$

we satisfy (3.6.4) for $k = 0$. Consider (3.6.4) for $k = 1$

$$(3.6.8) \qquad (\underset{\sim}{A}_o - \lambda_o)\underset{\sim}{u}_1 = (\lambda_1 - \underset{\sim}{A}_1)\underset{\sim}{u}_o .$$

This will not have a solution unless

$$(\underset{\sim}{\omega}_0, (\lambda_1 - \underset{\sim}{A}_1)\underset{\sim}{\omega}_0) = 0$$

which determines λ_1. With this choice of λ_1 we can write for $\underset{\sim}{u}_1$

$$\underset{\sim}{u}_1 = \alpha_1 \underset{\sim}{\omega}_o + \underset{\sim}{\omega}_1 \tag{3.6.9}$$

where $\underset{\sim}{\omega}_1$ denotes a particular solution to (3.6.8) and α_1 is an undetermined constant. Consider next

$$\begin{aligned}(\underset{\sim}{A}_0 - \lambda_0)\underset{\sim}{u}_2 &= (\lambda_1 - \underset{\sim}{A}_1)\underset{\sim}{u}_1 + (\lambda_2 - \underset{\sim}{A}_2)\underset{\sim}{u}_o \\ &= (\lambda_1 - \underset{\sim}{A}_1)\underset{\sim}{\omega}_1 + (\lambda_2 - \underset{\sim}{A}_2)\underset{\sim}{\omega}_0 + \alpha_1(\lambda_1 - \underset{\sim}{A}_1)\underset{\sim}{\omega}_o .\end{aligned} \tag{3.6.10}$$

Then the condition for solvability is

$$\begin{aligned}&(\tilde{\underset{\sim}{\omega}}_o, (\lambda_1 - \underset{\sim}{A}_1)(\underset{\sim}{\omega}_1 + \alpha_1 \underset{\sim}{\omega}_0) + (\lambda_2 - \underset{\sim}{A}_2)\underset{\sim}{\omega}_o) \\ &= (\tilde{\underset{\sim}{\omega}}_o, (\lambda_1 - \underset{\sim}{A}_1)\underset{\sim}{\omega}_1 + (\lambda_2 - \underset{\sim}{A}_2)\underset{\sim}{\omega}_o = 0\end{aligned} \tag{3.6.11}$$

determines λ_2 and we write the solution of (3.6.10) as

$$\underset{\sim}{u}_2 = \alpha_2 \underset{\sim}{\omega}_0 + \alpha_1 \underset{\sim}{\omega}_1 + \underset{\sim}{\omega}_2 . \tag{3.6.12}$$

The first term is of course a solution of the homogeneous equation, the second from (3.6.8) and (3.6.9) generates the last term of (3.6.10), and $\underset{\sim}{\omega}_2$ denotes a particular solution generating the first two terms of (3.6.10). For purposes of clarity, we consider one more step before writing out the general formulas. For $k = 3$ we have

$$(3.6.13)\qquad \begin{aligned}(\underset{\sim}{A}_0-\lambda_0)\underset{\sim}{u}_3 &= (\lambda_1-\underset{\sim}{A}_1)\underset{\sim}{u}_2 + (\lambda_2-\underset{\sim}{A}_2)\underset{\sim}{u}_1 + (\lambda_3-\underset{\sim}{A}_3)\underset{\sim}{u}_0\\ &= (\lambda_1-\underset{\sim}{A}_1)\underset{\sim}{\omega}_2 + (\lambda_2-\underset{\sim}{A}_2)\underset{\sim}{\omega}_1 + (\lambda_3-\underset{\sim}{A}_3)\underset{\sim}{\omega}_o\\ &+ \alpha_1\{(\lambda_1-\underset{\sim}{A}_1)\underset{\sim}{\omega}_1 + (\lambda_2-\underset{\sim}{A}_2)\underset{\sim}{\omega}_0\}\\ &+ \alpha_2(\lambda_1-\underset{\sim}{A}_1)\underset{\sim}{\omega}_o .\end{aligned}$$

Then from (3.6.11) λ_3 is determined from

$$(\tilde{\underset{\sim}{\omega}}_0, (\lambda_1-\underset{\sim}{A}_1)\underset{\sim}{\omega}_2 + (\lambda_2-\underset{\sim}{A}_2)\underset{\sim}{\omega}_1 + (\lambda_3-\underset{\sim}{A}_3)\underset{\sim}{\omega}_0) = 0$$

and letting $\underset{\sim}{\omega}_3$ be any particular solution of

$$(\underset{\sim}{A}_0-\lambda_0)\underset{\sim}{\omega}_3 = (\lambda_1-\underset{\sim}{A}_1)\underset{\sim}{\omega}_2 + (\lambda_2-\underset{\sim}{A}_2)\underset{\sim}{\omega}_1 + (\lambda_3-\underset{\sim}{A}_3)\underset{\sim}{\omega}_0$$

we have

$$\underset{\sim}{u}_3 = \alpha_3\underset{\sim}{\omega}_0 + \alpha_2\underset{\sim}{\omega}_1 + \alpha_1\underset{\sim}{\omega}_2 + \underset{\sim}{\omega}_3 .$$

Therefore, (3.6.4) is solved for any $k \leq r + 1$ by generating the λ_i and $\underset{\sim}{\omega}_i$ successively through

$$(3.6.14)\qquad (\underset{\sim}{\omega}_0, \sum_{i=1}^{k} (\lambda_i-\underset{\sim}{A}_i)\underset{\sim}{\omega}_{k-i}) = 0$$

$$(3.6.15)\qquad (\underset{\sim}{A}_0-\lambda_0)\underset{\sim}{\omega}_k = \sum_{i=1}^{k} (\lambda_i-\underset{\sim}{A}_i)\underset{\sim}{\omega}_{k-i}$$

and the solutions are given by

$$(3.6.16)\qquad \underset{\sim}{u}_k = \sum_{i=0}^{k} \alpha_i\underset{\sim}{\omega}_{k-i},\ \alpha_0 = 1.$$

It should be noted that although all the λ_i, $i = 0,\ldots,r + 1$ have been determined, no α_i, $i > 0$, has yet been found.

The determination of $\underset{\sim}{u}_{r+1}$ now follows from (3.6.5) for $k = r + 2$,

$$(\underset{\sim}{A}_o - \lambda_o)\underset{\sim}{u}_{r+2} = -\underset{\sim}{u}_1 + \sum_{i=1}^{r+2} (\lambda_i - \underset{\sim}{A}_i) \sum_{j=o}^{r+2-i} \alpha_j \underset{\sim}{\omega}_{r+2-i-j}$$

where we have substituted from (3.6.16) for $\underset{\sim}{u}_{r+2-i}$. Under an interchange in the orders of summation and some manipulation on the subscripts we get

$$\begin{aligned}(\underset{\sim}{A}_o - \lambda_o)\underset{\sim}{u}_{r+2} &= -(\underset{\sim}{\omega}_1 + \alpha_1 \underset{\sim}{\omega}_o) \\ &+ \sum_{j=1}^{r+2} \alpha_{r+2-j} \sum_{i=1}^{j} (\lambda_i - \underset{\sim}{A}_i)\underset{\sim}{\omega}_{j-i}.\end{aligned}$$

Multiplying on the left by $\tilde{\underset{\sim}{\omega}}_o$, the condition for solvability is

$$0 = -(\tilde{\underset{\sim}{\omega}}_o, \underset{\sim}{\omega}_1) - \alpha_1 (\tilde{\underset{\sim}{\omega}}_o, \underset{\sim}{\omega}_o) + \sum_{j=1}^{r+2} \alpha_{r+2-j} \sum_{i=1}^{j} (\tilde{\underset{\sim}{\omega}}_o, (\lambda_i - \underset{\sim}{A}_i)\underset{\sim}{\omega}_{j-i}).$$

But from (3.6.14, 15) the summation is zero for $j = 1,2,\ldots,r + 1$ and the condition is

$$\begin{aligned}0 &= (\tilde{\underset{\sim}{\omega}}_o, \underset{\sim}{\omega}_1) - \alpha_1 (\tilde{\underset{\sim}{\omega}}_o, \underset{\sim}{\omega}_o) \\ &+ \sum_{i=1}^{r+2} (\tilde{\underset{\sim}{\omega}}_o, (\lambda_i - \underset{\sim}{A}_i)\underset{\sim}{\omega}_{r+2-i}).\end{aligned}$$

We may take for convenience $(\tilde{\underset{\sim}{\omega}}_o, \underset{\sim}{\omega}_o) = 1$ and hence

(3.6.17) $$\alpha_1 = -(\tilde{\underset{\sim}{\omega}}_o, \underset{\sim}{\omega}_1) + \sum_{i=1}^{r+2} (\tilde{\underset{\sim}{\omega}}_o, (\lambda_i - \underset{\sim}{A}_i)\omega_{r+2-i}).$$

The procedure continues in this way - at each step we determine α_i and introduce the particular solution $\underset{\sim}{\omega}_{r+1+i}$.

This procedure is easily generalized to the case when $\underset{\sim}{A}_o$ has n linearly independent eigenvectors - but not necessarily n different eigenvalues. To see this, let us suppose λ_0 is an m-fold degenerate and $\underset{\sim}{\omega}_0^1, \underset{\sim}{\omega}_0^2, \ldots, \underset{\sim}{\omega}_0^m$ are m-linearly independent right eigenvectors and $\tilde{\underset{\sim}{\omega}}_0^1, \ldots, \tilde{\underset{\sim}{\omega}}_0^m$ m-linearly independent left eigenvectors. Our procedure is altered at first step since we now write

$$\underset{\sim}{u}_0 = \alpha_0^1 \underset{\sim}{\omega}_0^1 + \cdots + \alpha_0^m \underset{\sim}{\omega}_0^m$$

where the α_0^i, $i = 1,\ldots,m$ are unknown constants. At the second step we obtain

$$(\underset{\sim}{A}_0 - \lambda_0)\underset{\sim}{u}_1 = (\lambda_1 - \underset{\sim}{A}_1)(\alpha_0^1 \underset{\sim}{\omega}_0^1 + \cdots + \alpha_0^m \underset{\sim}{\omega}_0^m).$$

The condition for solvability now becomes

$$(\tilde{\underset{\sim}{\omega}}_0^i, (\lambda_1 - \underset{\sim}{A}_1)(\alpha_0^1 \underset{\sim}{\omega}_0^1 + \cdots + \alpha_0^m \underset{\sim}{\omega}_0^m)) = 0, \quad i = 1,\ldots,m.$$

This is a system of m-homogeneous equations in the unknowns $(\alpha_0^1, \ldots, \alpha^m)$ and hence in order to have a non-trivial solution the coefficient determinant must be zero. This gives us m determinations of λ_1. Choosing one of these the " eigenvector" $(\alpha_0^1, \ldots, \alpha_0^m)$ is determined (its magnitude may e.g., be set to unity). The calculation continues in this way. At each step we determine λ_k and $\alpha_k^1, \ldots, \alpha_k^m$ up to a (non-arbitrary) multiplicative constant. We do not give the details.

When $\underset{\sim}{A}_0$ does not have a full set of eigenvectors the problem becomes much more difficult. The main new feature is that fractional powers of z must be introduced. We shall discuss this more fully in the context of the scalar ordinary differential equation.

<u>Example</u>. For z large we consider

$$\frac{d\underset{\sim}{w}}{dz} = \begin{bmatrix} 1 & -\frac{1}{2z} \\ -\frac{1}{2z} & 0 \end{bmatrix} \underset{\sim}{w}. \tag{3.6.18}$$

According to the above notation, therefore

$$r = 0 \quad \underset{\sim}{A}_o = \begin{bmatrix} 1 & 0 \\ 0 & 0 \end{bmatrix}, \quad \underset{\sim}{A}_1 = \begin{bmatrix} 0 & -\frac{1}{2} \\ -\frac{1}{2} & 0 \end{bmatrix}$$

$\underset{\sim}{A}_k = 0$, $k > 1$. The eigenvalues and eigenvectors of $\underset{\sim}{A}_o$ are

$$\mu = \mu_o = 0, \qquad \underset{\sim}{u}_o^o = [0,1]$$
$$\mu = \mu_1 = 1, \qquad \underset{\sim}{u}_o^1 = [1,0].$$

We therefore seek solutions in the form

$$\underset{\sim}{w}^o = z^{\rho_o}[\underset{\sim}{u}_o^o + \frac{\underset{\sim}{u}_1^o}{z} + \cdots]$$
$$\underset{\sim}{w}^1 = e^z z^{\rho_1}[\underset{\sim}{u}_o^1 + \frac{\underset{\sim}{u}_1^1}{z} + \cdots].$$

Considering the $\underset{\sim}{w}^1$ calculation first, we find

$$\begin{bmatrix} 0 & 0 \\ 0 & -1 \end{bmatrix} \underset{\sim}{u}_k^1 = \begin{bmatrix} \rho_1 - k + 1 & \frac{1}{2} \\ \frac{1}{2} & \rho_1 - k + 1 \end{bmatrix} \underset{\sim}{u}_{k-1}^1.$$

For $k = 1$

$$\begin{bmatrix} 0 & 0 \\ 0 & -1 \end{bmatrix} \underset{\sim}{u}_1^1 = \begin{bmatrix} \rho_1 & \frac{1}{2} \\ \frac{1}{2} & \rho_1 \end{bmatrix} \begin{bmatrix} 1 \\ 0 \end{bmatrix}$$

Applying the condition for solvability we find

$$\rho_1 = 0.$$

In general, we can write solutions for $\underset{\sim}{u}_k^1$ as

$$\underset{\sim}{u}_k^1 = \alpha_k^1 \begin{bmatrix} 1 \\ 0 \end{bmatrix} + \omega_k^1 \begin{bmatrix} 0 \\ 1 \end{bmatrix} .$$

And for $k = 1$,

$$\underset{\sim}{u}_1^1 = \alpha_1^1 \begin{bmatrix} 1 \\ 0 \end{bmatrix} - \frac{1}{2} \begin{bmatrix} 0 \\ 1 \end{bmatrix} .$$

Continuing we have at the $k + 1$-step

$$\begin{bmatrix} 0 & 0 \\ 0 & -1 \end{bmatrix} \underset{\sim}{u}_{k+1}^1 = \begin{bmatrix} -k & \frac{1}{2} \\ \frac{1}{2} & -k \end{bmatrix} \left\{ \alpha_k^1 \begin{bmatrix} 1 \\ 0 \end{bmatrix} + \omega_k^1 \begin{bmatrix} 0 \\ 1 \end{bmatrix} \right\} .$$

The condition for solution then gives

$$\alpha_k^1 = \frac{\omega_k^1}{2k} .$$

But

$$\begin{aligned} \omega_{k+1}^1 &= k\omega_k^1 - \frac{\alpha_k^1}{2} \\ &= \omega_k^1 (k - \frac{1}{4k}) . \end{aligned}$$

Starting $\omega_1^1 = -\frac{1}{2}$ we can determine each ω_k^1 recursively and hence each α_k^1 and finally $\underset{\sim}{u}_k^1$. It is clear from the k that appears that the expansion grows like $k!$

Consider next the $\underset{\sim}{w}^0$ calculation

$$\begin{bmatrix} 1 & 0 \\ 0 & 0 \end{bmatrix} \underset{\sim}{u}_k^o = \begin{bmatrix} -k+1 & -\frac{1}{2} \\ -\frac{1}{2} & -k+1 \end{bmatrix} \underset{\sim}{u}_{k-1}^o$$

where $\rho_o = 0$ is easily seen to be the case.

The claim is that the solution for all k has the form

$$\underset{\sim}{u}_k^o = \alpha_k^o \begin{bmatrix} 0 \\ 1 \end{bmatrix} + \omega_k^o \begin{bmatrix} 1 \\ 0 \end{bmatrix}$$

In fact for $k = 1$

$$\underset{\sim}{u}_1^o = \alpha_1^o \begin{bmatrix} 0 \\ 1 \end{bmatrix} - \frac{1}{2} \begin{bmatrix} 1 \\ 0 \end{bmatrix} .$$

Assuming this we have

$$\begin{bmatrix} 1 & 0 \\ 0 & 0 \end{bmatrix} \underset{\sim}{u}_{k+1}^o = \begin{bmatrix} -k & -\frac{1}{2} \\ -\frac{1}{2} & -k \end{bmatrix} \left\{ \alpha_k^o \begin{bmatrix} 0 \\ 1 \end{bmatrix} + \omega_k^o \begin{bmatrix} 1 \\ 0 \end{bmatrix} \right\} .$$

The solvability condition is

$$\alpha_k^o = \frac{\omega_k^o}{2k}$$

and from this

$$\omega_{k+1}^o = \omega_k^o \left(\frac{1}{4} - k\right).$$

Then starting with $\omega_k^o = -\frac{1}{2}$ we may determine ω_i and α_i^o and hence $\underset{\sim}{u}_i^o$ for all i. An important point to note is that the series diverges. Hence even though (3.6.18) has a solution $z^p \underset{\sim}{u}(z)$ unlike the case of a regular singular point this is no guarantee that $\underset{\sim}{u}$ is analytic at infinity.

The situation depicted in the above example, i.e. of finding the expansion for $\underset{\sim}{w}$, to be divergent, should be regarded as typical. It is only in exceptional cases that the formal expansions for $\underset{\sim}{w}$ are convergent. There now arises the question of what meaning to attach to the formal solutions which we obtained. To answer this, let us consider a situation in which n independent formal solutions of (3.6.2) have been obtained,

$$(3.6.19) \qquad \underset{\sim}{w}^{\mu} = e^{q^{\mu}(z)} z^{\rho^{\mu}} [\underset{\sim}{u}_0 + \frac{\underset{\sim}{u}_1}{z} + \cdots].$$

Here the $q^{\mu}(z)$ are polynomials of degree $r + 1$ in z. From this we construct the formal fundamental matrix solution to (3.6.2) in the form

$$(3.6.20) \qquad \underset{\sim}{W} = \underset{\sim}{U} z^{\underset{\sim}{P}} e^{\underset{\sim}{Q}}$$

where $\underset{\sim}{Q}$ is diagonal with elements $q^1,\ldots,q^n$ and $\underset{\sim}{P}$ is diagonal with $\rho^1,\ldots,\rho^n$. The columns of $\underset{\sim}{U}$ are the formal expansions

$$[\underset{\sim}{u}_0^i + \frac{\underset{\sim}{u}_1^i}{z} + \cdots], \quad i = 1,\ldots,n.$$

The basic content of the existence theorems to be found in [7] and [10] is that there exists a fundamental matrix solution to (3.6.2), $\underset{\sim}{\tilde{W}}$ such that for $z \to \infty$, and some sector of the complex plane,

$$\underset{\sim}{\tilde{W}} \sim \underset{\sim}{W}.$$

To understand more fully the role of the sector in this statement we note that the asymptotic behavior of (3.6.19) changes across the lines

$$\text{Re } q^{\mu} = 0$$

called edges. For $z \to \infty$ the edges are determined by

$$\mathrm{Re}(\lambda_o z^{r+1}) = 0.$$

Writing

$$\lambda_o = |\lambda_o| e^{i\theta_o}$$

and

$$z = |z| e^{i\theta}$$

the asymptotes of the edges are given by

$$\theta = -\frac{\theta_o}{r+1} + \frac{(2i+1)\pi}{2(r+1)}, \quad i = 0, \pm 1, \ldots . \tag{3.6.21}$$

When $\mathrm{Re}\ q^{\mu} > 0$ the solution (3.6.19) is said to be dominant and when $\mathrm{Re}\ q^{\mu} < 0$ it is said to be sub-dominant. Hence (3.6.19) is dominant or sub-dominant (recessive) in sectors of size

$$\frac{\pi}{r+1} .$$

Next, it is clear that the rays are Stokes lines. For suppose $\tilde{\underset{\sim}{w}}$ is a solution to (3.6.2) such that

$$\tilde{\underset{\sim}{w}} \sim \underset{\sim}{w}^{\mu}$$

in a sector in which $\underset{\sim}{w}^{\mu}$ is dominant. Then on passing through the edge of $\underset{\sim}{w}^{\mu}$, it becomes recessive. There are now countless ways in which the true solution $\tilde{\underset{\sim}{w}}^{\mu}$ can behave.

The theorem of existence states that if $\underset{\sim}{A}$ is analytic at infinity and for any sector S less than $\pi/r+1$, there exists a fundamental matrix $\underset{\sim}{X}$ such that

$$\underset{\sim}{X} \sim \underset{\sim}{W}\underset{\sim}{C}.$$

For some constant matrix $\underset{\sim}{C}$.

On passing out of this sector there will be a new matrix $\tilde{\underset{\sim}{C}}$ such that

$$\underset{\sim}{X} \sim \underset{\sim}{W}\tilde{\underset{\sim}{C}}.$$

The connection between $\tilde{\underset{\sim}{C}}$ and $\underset{\sim}{C}$ is not a priori clear.

Finally, even if $\underset{\sim}{A}$ is not analytic in the neighborhood of infinity but has an asymptotic expansion (3.6.3) in some sector, the theorem remains valid now, however, some care is necessary in the discussion of the sectors (see [7]).

In the general case, when no special properties of $\underset{\sim}{A}_0$ are demanded, a formal fundamental matrix of the form (3.6.20) may again be constructed. $\underset{\sim}{Q}$ is diagonal, but its entries are polynomials in $z^{1/n}$ when n is an integer. $\underset{\sim}{P}$ is constant but not necessarily diagonal and the columns of $\underset{\sim}{U}$ are formal expansions in inverse powers of $z^{1/n}$. Finally the fundamental existence theorem states that for each sufficiently small sector and $z \to \infty$ a fundamental matrix solution exists which has (3.6.20) as its asymptotic development (see [7]).

When fraction powers enters (3.6.20) is called subnormal otherwise it is said to be normal.

As a general illustration of the theory discussed above, we consider

$$\frac{d\underset{\sim}{w}}{dz} = \underset{\sim}{A}_0\underset{\sim}{w} + \frac{1}{z}\underset{\sim}{A}_1\underset{\sim}{w} \tag{3.6.22}$$

where $\underset{\sim}{A}_0, \underset{\sim}{A}_1$ are constant $n \times n$ matrices. Also for further simplicity we take $\underset{\sim}{A}_0$ as being diagonal with distinct eigenvalues. We integrate (3.6.22) using the method of Laplace, i.e., we seek a solution in the form

$$(3.6.23) \qquad \underset{\sim}{w} = \int_P e^{tz} \underset{\sim}{\phi}(t)dt$$

where for the moment the path of integration P and the vector $\underset{\sim}{\phi}(t)$ are known.

By formal substitution of (3.6.23) into (3.6.21) we obtain

$$(3.6.24) \qquad \begin{aligned} \left(\frac{d}{dz} - \underset{\sim}{A}_o - \frac{\underset{\sim}{A}_1}{z}\right)\underset{\sim}{w} &= \int_P e^{tz}\left[t - \underset{\sim}{A}_o - \frac{\underset{\sim}{A}_1}{z}\right]\underset{\sim}{\phi}(t)dt \\ &= z^{-1} \int_P e^{tz}[z(t-\underset{\sim}{A}_o) - \underset{\sim}{A}_1]\underset{\sim}{\phi}(t)dt. \end{aligned}$$

Next consider

$$\int_P e^{tz} z(t-\underset{\sim}{A}_o)\underset{\sim}{\phi}(t)dt = \int_P (t-\underset{\sim}{A}_o)\underset{\sim}{\phi}(t) \frac{d}{dt} (e^{tz})dt$$

$$= \int_P \frac{d}{dt} \{(t-\underset{\sim}{A}_o)\underset{\sim}{\phi}(t)e^{tz}\}dt - \int_P e^{tz}\{(t-\underset{\sim}{A}_o)\underset{\sim}{\phi}'(t) + \underset{\sim}{\phi}(t)\}dt$$

$$= [(t-\underset{\sim}{A}_o)\underset{\sim}{\phi}(t)e^{tz}]_P - \int_P e^{tz}\{(t-\underset{\sim}{A}_o)\underset{\sim}{\phi}'(t) + \underset{\sim}{\phi}(t)\}dt$$

where the first term represents the difference of the integrand taken at the end-points of P. Substituting into (3.6.22), we obtain

$$\begin{aligned} \left(\frac{d}{dz} - \underset{\sim}{A}_o - \frac{\underset{\sim}{A}_1}{z}\right)\underset{\sim}{w} &= z^{-1}[(t-\underset{\sim}{A}_o)\underset{\sim}{\phi}(t)e^{tz}]_P \\ &\quad - z^{-1} \int_P e^{tz}\{(t-\underset{\sim}{A}_o)\underset{\sim}{\phi}'(t) + (\underset{\sim}{A}_1+\underset{\sim}{I})\underset{\sim}{\phi}\}dt. \end{aligned}$$

Hence (3.6.23) satisfies (3.6.22) if $\underset{\sim}{\phi}(t)$ is such that

$$(3.6.25) \qquad [(t-\underset{\sim}{A}_o)\underset{\sim}{\phi}(t)e^{tz}]_P = 0$$

and

$$(3.6.26)\qquad (t-\underset{\sim}{A}_0)\frac{d}{dt}\underset{\sim}{\Phi}(t) = -(\underset{\sim}{A}_1+\underset{\sim}{I})\underset{\sim}{\Phi}.$$

The system of linear ordinary differential equations generated in this way is of the same order as the original system (3.6.22), however, the new system (3.6.25) now has only regular singular points, including the point at infinity. We may, therefore, use the methods of section 3.5 to analyze (3.6.26).

Let us represent then entries of $\underset{\sim}{A}_0$ (diagonal) by a_i, $i = 1,\ldots,n$. Then at each point

$$t = a_i, \qquad i = 1,\ldots,n$$

and $t = \infty$, the system has a regular singular point. Therefore, writing

$$\underset{\sim}{\Phi}(t) = (t-a_i)^{\rho_i}\underset{\sim}{\psi}^i(t) = (t-a_i)^{\rho_i}[\underset{\sim}{\psi}_o^i + (t-a_i)\underset{\sim}{\psi}_1^i + \cdots]$$

where $\underset{\sim}{\psi}$ is analytic at $t = a_i$, we find,

$$\begin{aligned}(t-\underset{\sim}{A}_o)[\rho_i(t-a_i)^{\rho_i-1}&\{\psi_o^i + (t-a_i)\underset{\sim}{\psi}_1^i + \cdots\}\\ &+ (t-a_i)^{\rho_i}\{\underset{\sim}{\psi}_1^i + \cdots\}]\\ &= -(\underset{\sim}{A}_1+\underset{\sim}{I})\{(t-a_i)^{\rho_i}[\underset{\sim}{\psi}_o^i + (t-a_i)\underset{\sim}{\psi}_1^o + \cdots]\}.\end{aligned}$$

Therefore, there is just one root unequal to zero and this is

$$\rho_i = -(1 + \alpha_{ii})$$

where

$$\alpha_{ii} = (A_1)_{ii}.$$

On choosing the n solutions of the form

$$\underset{\sim}{\phi} = (t-a_i)^{-1-\alpha_{ii}} \underset{\sim}{\psi}_i$$

we generate n linearly independent solutions. If ρ_i is positive we can take P in (3.6.23) to extend from $t = a_i$ to infinity and the path at ∞ is such that

$$\frac{\pi}{2} < \arg z + \arg t < \frac{3\pi}{2}. \tag{3.6.27}$$

If ρ_i is zero (multiple eigenvalues, Case 1, Section 3.5) or a negative integer (eigenvalues differing by an integer, Case 2, Section 3.5) the vector $\underset{\sim}{\psi}^i(t)$ will include a logarithmic singularity at $t = a_i$. Also for other negative values of ρ_i, $t = a_i$ is a branch point of $\underset{\sim}{\phi}$. In these cases we take P to "loop" $t = a_i$ coming from a direction given by (3.6.27) and then returning to infinity in the same direction. In all cases the paths P_i must avoid the $a_j \neq a_i$ since these are in general singularities of $\underset{\sim}{\psi}_i(t)$. In this way we generate n solutions of (3.6.22) in the forms

$$\underset{\sim}{w}^i = \int_{P_i} e^{tz}(t-a_i)^{\rho_i} \underset{\sim}{\psi}^i(t)dt. \tag{3.6.28}$$

Each of these integrals is in the form of a Laplace integral and its asymptotic expansion for $|z| \to \infty$ follow from the expansion of $\underset{\sim}{\psi}^i(t)$ at $t = a_i$. The results will of course be the same as gotten earlier in this section by more direct means.

Although $\underset{\sim}{\psi}^i(t)$ is analytic in the neighborhood of $t = a_i$ this may no longer be true at the other singular points $t = a_j$, $j \neq i$. Next, we analytically continue each solution of the form (3.6.28). We follow the same procedure given in Section 2.1. As the argument of z changes we accordingly rotate the path of integration, P_i. In doing this we encounter other a_j which are singularities of the integrand. Typically, we obtain a distorted path as shown in the Figures 1 - 3.

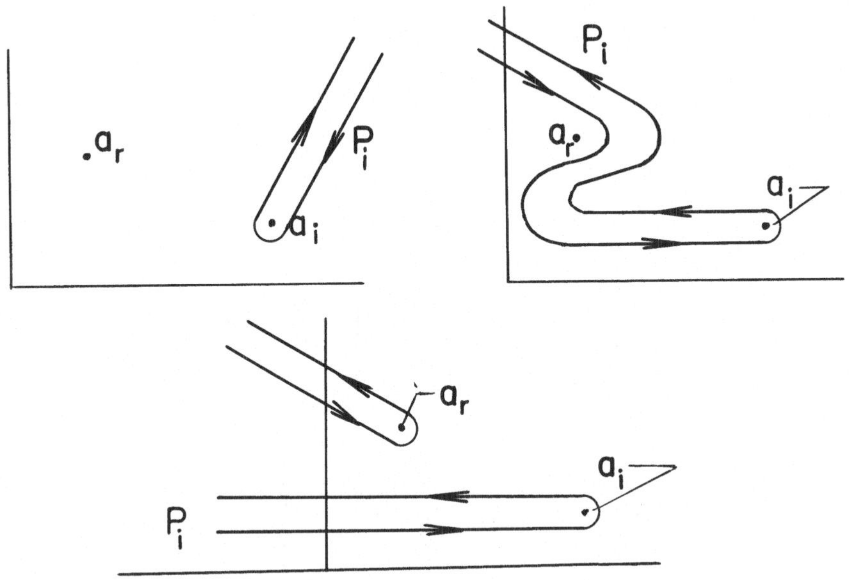

In going from Figure 2 to Figure 3 we combine the contributions from the two paths looping a_r. Since as drawn, a_i is a branch point, there is no cancellation. In this way the presence of a Stokes is illustrated.

<u>Example</u>. Consider

$$\frac{d\underset{\sim}{w}}{dz} = \begin{bmatrix} 1 & -\frac{1}{2z} \\ -\frac{1}{2z} & -\frac{1}{2z} \end{bmatrix} \underset{\sim}{w}.$$

Applying the above discussion, the differential equation governing $\underset{\sim}{\phi}$ is

$$\begin{bmatrix} t-1 & 0 \\ 0 & t \end{bmatrix} \frac{d}{dt} \underset{\sim}{\phi} = \begin{bmatrix} -1 & \frac{1}{2} \\ \frac{1}{2} & -\frac{1}{2} \end{bmatrix} \underset{\sim}{\phi}$$

The equation has regular singular points at $0, 1, \infty$. We first consider $\underset{\sim}{\phi}$ in the neighborhood of the origin and denoting it by $\underset{\sim}{\phi}^{o}$

$$t \frac{d\underset{\sim}{\phi}^{o}}{dt} = \begin{bmatrix} \frac{1}{t-1} & 0 \\ 0 & 1 \end{bmatrix} \begin{bmatrix} -1 & \frac{1}{2} \\ \frac{1}{2} & -\frac{1}{2} \end{bmatrix} \underset{\sim}{\phi}^{o} = \begin{bmatrix} \frac{t}{1-t} & \frac{t}{2(t-1)} \\ \frac{1}{2} & -\frac{1}{2} \end{bmatrix} \underset{\sim}{\phi}^{o}$$

$$= \left\{ \begin{bmatrix} 0 & 0 \\ \frac{1}{2} & -\frac{1}{2} \end{bmatrix} + t \begin{bmatrix} 1 & -\frac{1}{2} \\ 0 & 0 \end{bmatrix} + \cdots \right\} \underset{\sim}{\phi}^{o} .$$

And on writing

$$\underset{\sim}{\phi}^{o} = t^{-1/2} \underset{\sim}{\psi}_{o} = t^{-1/2} [\underset{\sim}{\psi}_{o}^{o} + t\underset{\sim}{\psi}_{o}' + \cdots]$$

we obtain

$$\underset{\sim}{\psi}_{o}^{o} = \begin{bmatrix} 0 \\ 1 \end{bmatrix}, \quad \underset{\sim}{\psi}_{o}' = -\begin{bmatrix} 1 \\ \frac{1}{2} \end{bmatrix}, \ldots .$$

The solution to $\underset{\sim}{w}$ corresponding to this is

$$\underset{\sim}{w}^{o} = \int_{P_o} e^{tz} \frac{\underset{\sim}{\psi}(t)}{\sqrt{t}} dt$$

Suppose $\operatorname{Re} z > 0$ and take the path P to start from $-\infty - i\epsilon$ loop the origin and then go to $-\infty + i\epsilon$ $(\epsilon > 0)$. By elementary changes of variable

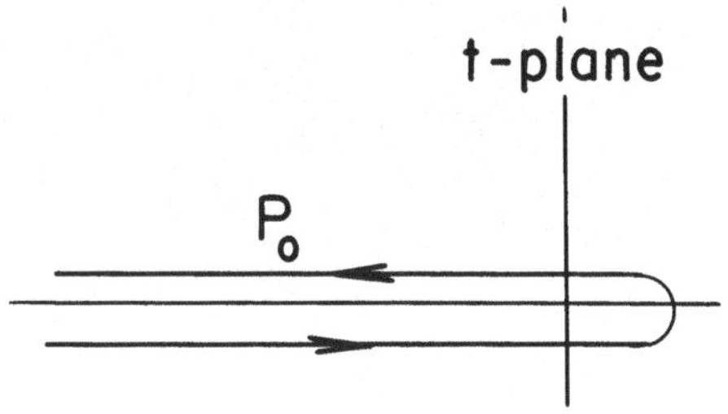

$$\underset{\sim}{w}^{0} = 2i \int_0^\infty e^{-sz} \frac{\underset{\sim}{\psi}^{0}(-s)}{\sqrt{s}} ds .$$

We next seek the solution in neighborhood of $t = 1$. This we denote by $\bar{\underset{\sim}{\varphi}}'$, so that

$$(t-1) \frac{d}{dt} \bar{\underset{\sim}{\varphi}}' = \begin{bmatrix} 1 & \frac{1}{2} \\ \frac{t-1}{t} & -\frac{t-1}{2t} \end{bmatrix} \bar{\underset{\sim}{\varphi}}'$$

where the differential equation has been rewritten in a suitable form. In this case if we expand in the neighborhood of $t = 1$ we find,

$$A_o = \begin{bmatrix} -1 & \frac{1}{2} \\ 0 & 0 \end{bmatrix}$$

so that the eigenvalues differ by an integer. Therefore, Case 2 of Section 3.5 applies, and we write

$$\bar{\underset{\sim}{\varphi}}' = z^{\lambda}(\lambda+1)[\underset{\sim}{\psi}_1^{0} + z\underset{\sim}{\psi}_1^{1} + \cdots].$$

The solution in question is

$$\bar{\underset{\sim}{\varphi}}' = \left.\frac{d\bar{\underset{\sim}{\varphi}}'}{d\lambda}\right|_{\lambda=-1} .$$

Leaving the details as an exercise we state the result,

$$\mathfrak{v}' = \frac{1}{t-1}\begin{bmatrix} 1 \\ \\ 0 \end{bmatrix} + \ln(t-1)\begin{bmatrix} -\frac{1}{4} \\ \\ \frac{1}{2} \end{bmatrix} + \begin{bmatrix} -\frac{1}{4} \\ \\ 0 \end{bmatrix} + \cdots .$$

Again supposing that $\operatorname{Re} z > 0$ we take

$$\underset{\sim}{w}' = \int_{P_1} e^{tz}\mathfrak{v}'(t)dt$$

where the path P_1 is shown in the figure.

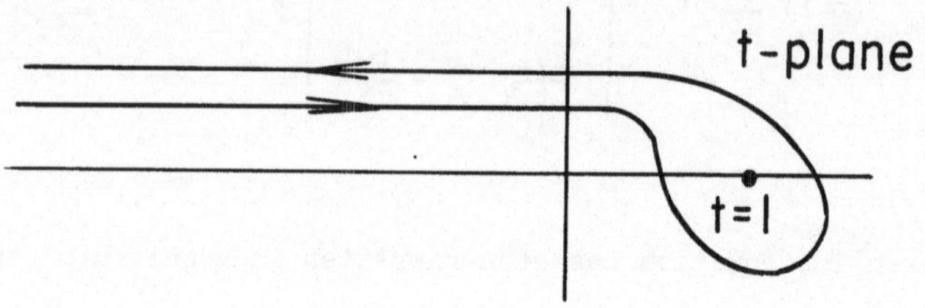

Scalar Ordinary Differential Equations

Consider

$$p_0 \frac{d^n w}{dz^n} + p_1(z)\frac{d^{n-1}w}{dz^{n-1}} + p_2(z)\frac{d^{n-2}w}{dz^{n-2}} + \cdots + p_n w = 0 \tag{3.6.29}$$

with $p_0 = 1$. As in the matrix case we restrict attention to the neighborhood of infinity. The coefficients in the neighborhood of infinity are assumed to have the expansions

$$p_j \sim z^{\gamma_j}[a_{j0} + \frac{a_{j1}}{z} + \frac{a_{j2}}{z^2} + \cdots], \qquad j = 1,\ldots,n. \tag{3.6.30}$$

Since (3.6.29) is assumed to have ∞ as irregular singular point, at least one of the integer exponents γ_j is such that

$$\gamma_j \geq 1 - j.$$

With the matrix case as motivation, we write the solution in the form

$$(3.6.31) \qquad \begin{aligned} w &= e^{q(z)} \\ q &= \frac{\alpha z^s}{s} + r(z) \\ r(z) &= o(z^s) \end{aligned}$$

For the moment, we seek solutions such that $s > 0$. Then writing

$$(3.6.32) \qquad \frac{d^i w}{dz^i} = \tau_i e^q$$

we see that

$$\tau_0 = 1, \quad \tau_1 = q', \ldots, \tau_{j+1} = \tau_j' + \tau_j q'$$

which directly gives the estimate

$$(3.6.33) \qquad \tau_j = O(z^{j(s-1)}).$$

Equation (3.6.29) becomes,

$$(3.6.34) \qquad p_0 \tau_n + p_1 \tau_{n-1} + \cdots + p_n \tau_0 = 0.$$

Then from (3.6.30) and (3.6.33)

$$p_j \tau_{n-j} = O(z^{\gamma_j + (n-j)(s-1)}), \qquad j = 0, \ldots, n.$$

The successive terms in (3.6.34), therefore have leading exponents

(3.6.35) $$e_j = \gamma_j + (n-j)(s-1), \quad j = 0,\dots,n$$

with $\gamma_0 = 0$. In order for there to be a solution in the form (3.6.31), it is necessary for s to be such that no e_j be greater than all other exponents. As a direct approach we can set two of these equal to one another and find a value of s from this. E.g., suppose s_{pq} is such that

$$e_p(s_{pq}) = e_q(s_{pq})$$

then it is an acceptable exponent only if

$$e_p(s_{pq}) \geq e_j(s_{pq}), \quad j = 0,\dots,n.$$

There is, however, a geometrical construction - known as Newton's polygon - for the systematic solution of all eligible values of s.

Consider the $n + 1$ lattice points

$$(0,n),\ (\gamma_1,n-1),\dots,(\gamma_j,n-j),\dots,(\gamma_n,0)$$

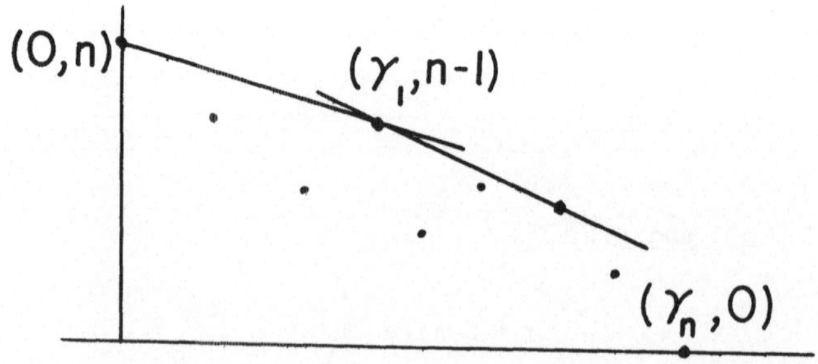

Consider a ray through $(0,n)$ and parallel to the x-axis. This is rotated clockwise until it meets one or more of the plotted lattice points. The same process is repeated using the most remote of the encountered points. This process is repeated until reaching $(\gamma_n,0)$ or the slope, $\frac{1}{\mu}$, becomes infinite - whichever comes first. Note that all lattice points always lie below or on the lines of the polygon.

Suppose $(\gamma_p,n-p)$, $(\gamma_q,n-q)$ determined one of these sides of the polygon. Further let

$$x + \mu y = c$$

denote the equation joining these points. Then

$$e_q = \gamma_q + \mu(n-q) = \gamma_p + \mu(n-p) = e_p = c.$$

Moreover, it is clear from the construction that if $(\gamma_j,(n-j))$ is not on this line then

$$e_p > \gamma_j + \mu(n-j).$$

Hence

$$s = \mu + 1$$

is an admissible value and hence all admissible values are obtained in this way. We notice that from this calculation there is no guarantee that s is an integer and, in fact, all that can be said of s is that it is rational.

It is next necessary to calculate the coefficient α and more generally the expansion of the remainder term $r(z)$ of $q(z)$, (3.6.31). Turning first to the calculation of α, let s_0 denote an admissible value. Generally speaking we will have,

$$e_{\nu_1}(s_o) = e_{\nu_2}(s_o) = \cdots = e_{\nu_k}(s_o).$$

To be definite we write,

$$\nu_1 < \nu_2 < \cdots < \nu_k.$$

With $s = s_o$ the leading term of (3.6.34) is

$$z^{e_{\nu_1}(s_o)}\left[a_{\nu_1 o}\alpha^{n-\nu_1} + a_{\nu_2 o}\alpha^{n-\nu_2} + \cdots + a_{\nu_k o}\alpha^{n-\nu_k}\right].$$

Hence there are $\nu_k - \nu_j$ values of α, corresponding to s_o, and determined by

$$a_{\nu_1 o}\alpha^{\nu_k-\nu_j} + \cdots + a_{\nu_k o} = 0. \tag{3.6.36}$$

To determine $r(z)$ we may write

$$r(z) = \frac{\beta z^t}{t} + o(z^t)$$

and proceed as before. However, with α and s_o known, certain shortcuts may be taken to simplify the calculation. First suppose s, admissible, in lowest terms has the form

$$s = \frac{m}{p},$$

with p and m integer. This signals an expansion of $q(z)$, in powers of $z^{1/p}$ In particular we can write,

$$\begin{aligned} q(z) = \frac{p\alpha_m z^{m/p}}{m} + \frac{p\alpha_{m-1}}{m-1} z^{(m-1)/p} + \cdots + p\alpha_2 z^{1/p} + \alpha_1 \ln z \\ + p\alpha_o z^{-1/p} + \cdots . \end{aligned} \tag{3.6.37}$$

Actually (3.6.37) will not be adequate in the event that α_m, the solution of (3.6.36), is not a simple root. If M is the multiplicity of α_m in (3.6.36) and if $s = m/p$ is the admissible value of s, then an expansion in $z^{1/pM}$ must be sought, i.e.

$$q(z) = pM\frac{\alpha_{mM} z^{mM/pM}}{mM} + \frac{pMz^{(mM-1)/pM}}{mM-1} + \cdots + pM\alpha_2 z^{1/pM} + \alpha_1 \ln z \tag{3.6.38}$$
$$+ pM\alpha_0 z^{-1/pM} + \cdots .$$

Hence even if $p = 1$, i.e., s integer, an expansion in fractional powers of α may be necessary if α is not a simple root of (3.6.36). In a moment we will illustrate these points in an example.

In addition to the above solutions we can also have those of the type,

$$w = z^\rho[u_0 + \frac{u_1}{z} + \cdots]. \tag{3.6.39}$$

The formal procedure now follows that used in the discussion of regular singular points. It should be noted, however, that if (3.6.39) is substituted into (3.6.29) the leading exponents of z of each term have the forms,

$$\gamma_0 + \rho - n,\ \gamma_1 + (\rho-n+1),\ldots,\gamma_n + \rho.$$

Therefore, the polygon method used above, can also be used in the determination of ρ.

The task of enumerating the various formal solutions to show that n are obtained is tedious and we skip this step. (An excellent source for this material is the book by Ince [16]).

<u>Example.</u> Consider

$$\frac{d^2w}{dz^2} + (z + \frac{1}{z})\frac{dw}{dz} + w = 0.$$

Here $\gamma_1 = 0 = \gamma_2$ and therefore $s = 1$, is the only admissible value. Writing

$$w = e^q$$

$$q = \alpha z + r(z)$$

we get

$$(\alpha^2+2\alpha+1) + 2r'(\alpha+1) + (r')^2 + \frac{\alpha}{z} + \frac{r'}{z} = 0.$$

Therefore, $\alpha = -1$ is a double root. According to the above discussion (corresponding to (3.6.38)) we write

$$w = e^q = \exp[\alpha z + \beta z^{\frac{1}{2}} + \gamma \ln z + \frac{\delta}{z^{\frac{1}{2}}} + \cdots].$$

Substituting we easily find

$$\alpha = -1,\ \beta = \pm 2,\ \gamma = -\frac{1}{4}, \ldots .$$

Second Order Equations

For the case of second order equations many of the technical problems may be circumvented by a series of transformations. Consider

$$\frac{d^2w}{dz^2} + p(z)\frac{dw}{dz} + q(z)w = 0.$$

Then setting

$$w = e^{-\frac{1}{2}\int^z p\,dz}\, v$$

(where the integral represents an indefinite integral) we obtain,

$$\frac{d^2v}{dz^2} + (q - \frac{p'}{2} - \frac{p^2}{4})v = 0.$$

Next, we set

$$z = z^2$$
$$v = z^{1/2}u$$

to obtain

$$\frac{d^2u}{dz^2} + [4z^2(q - \frac{p'}{2} - \frac{p^2}{4}) - \frac{3}{4z^2}]u = 0.$$

This second transformation has been introduced in order to avoid the appearance of fractional powers.

We may, therefore, consider without loss of generality,

(3.6.40)
$$\frac{d^2w}{dz^2} + z^{2r}q(z)w = 0$$

and take

$$q(z) \sim [q_0 + \frac{q_1}{z} + \cdots]$$

with the integer $r \geq 0$, and $q_0 \neq 0$. To find the asymptotic forms of the solutions, we write formally

(3.6.41)
$$w = e^{s(z)}$$

with

(3.6.42)
$$s(z) = \frac{s_0 z^{r+1}}{r+1} + \frac{s_1 z^r}{r} + \cdots + s_{r+1} \ln z + \frac{s_{r+2}}{z} + \cdots$$

[Actually it is customary to write instead $e^{\alpha(z)}z^{\rho}U(z)$ with α a polynomial of degree $r + 1$, ρ a constant and $U(z)$ a formal Laurent expansion. We bypass the formal identification of the two approaches.]

On substituting (3.6.41) into the equation (3.6.40) we have

$$(s')^2 + s'' + z^{2r}q = 0.$$

Substituting (3.6.42) into this equation and setting the coefficients of z^{2r-k} to zero

$$\sum_{i+j=k} s_i s_j + q_k = 0, \quad k = 0,\ldots,r$$

$$\sum_{i+j=k+r+1} s_i s_j + (r-k)s_k + q_{r+1+k} = 0, \quad k = 0,1,\ldots .$$

The solutions are

$$s_o = \pm\sqrt{-q_o}$$

(3.6.43) $$s_k = -\frac{1}{2s_o}[q_k + \sum_{i=1}^{k-1} s_i s_{k-i}], \quad k = 1,0,\ldots,r$$

$$s_k = -\frac{1}{2s_o}[q_k + \sum_{i=1}^{k-1} s_i s_{k-i} - (2r-k+1)s_{k-r-1}], \quad k = r+1,\ldots,\infty.$$

Example. Bessel's equation

$$\frac{d^2w}{dz^2} + \frac{1}{z}\frac{dw}{dz} + (1 - \frac{n^2}{z^2})w = 0.$$

Writing

$$w = \frac{v}{\sqrt{z}}$$

we obtain

$$\frac{d^2v}{dz^2} + (1 + \frac{1-4n^2}{4z^2})v = 0.$$

Next introducing the formal expansion

$$v = e^{s(z)}$$

with

$$s(z) = s_o z + s_1 \ln z + \frac{s_2}{z} + \cdots$$

since $r = 0$, also

$$q_o = 1,\ q_1 = 0,\ q_2 = \frac{1-4n^2}{4},\ q_k = 0, \qquad k > 2.$$

Introducing these into (3.6.43) we have

$$s_o = \pm i$$

$$s_1 = \pm \frac{i}{2} [q_1 = 0] = 0$$

$$s_2 = \pm \frac{i}{2} [\frac{1-4n^2}{4}].$$

It is customary to write the asymptotic expansions as

$$v = e^{\pm iz}[1 \pm \frac{(1-4n^2)i}{8z} + \cdots\cdot]$$

$$w = \frac{e^{\pm iz}}{\sqrt{z}} [1 \pm \frac{(1-4n^2)i}{8z} + \cdots\cdot] .$$

Example. Airy's Equation

$$\frac{d^2w}{dz^2} - zw = 0.$$

Introducing $z = z^2$ and $w = z^{1/2}v$, then

$$\frac{d^2v}{dz^2} + z^4(-4 + \frac{1}{4z^6})v = 0.$$

Therefore,

$$r = 2$$
$$q_0 = -4$$
$$q_6 = \frac{1}{4}$$
$$q_k = 0, \ k \neq 0,6.$$

The calculations are straightforward and we find,

$$v \sim \frac{e^{\pm\frac{2}{3}z^3}}{z}[1 + \cdots]$$

or

$$w \sim \frac{e^{\pm\frac{2}{3}z^{3/2}}}{z^{1/4}}[1 + \cdots].$$

Exercise 84. Find the asymptotic developments for solutions of

$$w'' - zw = 0$$

when $z \to \infty$ and compare with the results of Section 2.7.

Exercise 85. Consider solutions of the Hamburger equation,

$$w'' - (a + \frac{2b}{z} + \frac{c}{z^2})w = 0$$

for $z \to \infty$. In particular, discuss their various forms with different choices of the constants a,b,c.

3.7. Ordinary Differential Equations Containing a Large Parameter.

We focus on the case of a second order ordinary differential equation depending on a parameter λ,

(3.7.1) $$w'' + q(x,\lambda)w = 0.$$

(Recall that the transformation $u = w \exp[-\frac{1}{2}\int^{t} pdz]$ transforms $u'' + pu' + u = 0$ into the form (3.7.1)). Equations of the type (3.7.1) occur most frequently in practice. An extensive treatment of the general matrix case is given in the book by Wasow [7].

We assume that $q(x; \lambda)$ has the expansion

(3.7.2) $$q \sim \sum_{n=0} q_n(x)\lambda^{2k-n}$$

for $n \to \infty$, and

(3.7.3) $$q_0(x) \neq 0$$

at least in the region of interest. There is no loss of generality in assuming that the lead term is to an even power in λ, since merely written $\lambda = \mu^2$ effects this form. Finally, for convenience, we restrict attention to the case of real x. As will be clear if $q(x; \lambda)$ is complex analytic in x the formal procedures remain the same in the complex plane.

Formal Solution.

We formally solve (3.7.1) subject to (3.7,2,3) by writing

(3.7.4) $$\begin{cases} w = e^{\beta(x,\lambda)} \\ \beta(x;\lambda) = \sum\limits_{n=o} \lambda^{k-n}\beta_n(x) \end{cases}$$

This choice of the assumed expansion is immediately suggested by the constant coefficient case. Substituting (3.7.4) into (3.7.1) we obtain,

$$\beta'' + (\beta')^2 + q = 0$$

and from (3.7.2) and (3.7.4)

$$\sum_{n=o}^{\infty} \lambda^{k-n}\beta_n'' + \sum_{m=o}^{\infty} \lambda^{2k-m} \sum_{i+l=m} \beta_i'\beta_j' + \sum_{m=o}^{\infty} \lambda^{2k-m} q_m = 0.$$

Therefore,

$$q_m + \sum_{i=o}^{m} \beta_i'\beta_{m-i}' = 0, \quad m = 0,\ldots,k-1$$

$$\beta_{n-k}'' + q_m + \sum_{i=o}^{m} \beta_i'\beta_{m-i}' = 0, \quad m > k.$$

In particular,

(3.7.5) $$(\beta_o')^2 + q_o = 0 \quad \text{or}$$
$$\beta_o = \pm i \int \sqrt{q_o}\, dx$$

and in general

(3.7.6) $$\begin{cases} \beta_m' = \pm \dfrac{i}{2\sqrt{q_o}} \left(\sum\limits_{i=1}^{m-1} \beta_i'\beta_{m-i}' - q_m \right), & m = 1,\ldots,k-1 \\ \beta_m' = \mp \dfrac{i}{2\sqrt{q_o}} \left(\sum\limits_{i=1}^{m-1} \beta_i'\beta_{m-i}' - q_m - \beta_{m-k}'' \right), & m \geq k \end{cases}$$

Example. The most common form of (3.7.1) is

$$\frac{d^2u}{dx^2}u + \lambda^2 f(x)u = 0. \tag{3.7.7}$$

In this case $k = 1$, $q_0 = f$, $q_k = 0$, $k > 0$. The equation for β_1, from (3.7.6), is

$$\beta_1' = \frac{\pm i}{2\sqrt{q_0}} \beta_0'' = \frac{\pm i}{2\sqrt{q_0}} \frac{d}{dx} \quad \frac{\pm i q_0'}{2\sqrt{q_0}}$$

and hence

$$\beta_1 = -\ln q_0^{1/4} + \text{const.}$$

From this we get

$$u \sim \frac{1}{(q_0)^{1/4}} e^{\pm i\lambda \int^x \sqrt{q_0}(s)\, ds}$$

These asymptotic solutions are sometimes referred to as the WKB solutions.

Exercise 86. Develop the formal theory by assuming

$$w = [w_0 + \frac{w_1}{\lambda} + \frac{w_2}{\lambda^2} + \dots]\exp[\beta_0\lambda^k + \dots \beta_{k-1}\lambda]$$

instead of (3.7.4).

Exercise 87. Prove that the WKB solutions are exact when

$$f(x) = \frac{1}{[Cx+D]^4}$$

in (3.7.7).

Turning or Transition Points.

We have formally demonstrated that

$$\frac{d^2u}{dx^2} + \lambda^2 f(x)u = 0 \tag{3.7.7}$$

has the asymptotic or WKB solutions

$$\frac{1}{[f]^{1/4}} \exp [\pm i\lambda \int^x \sqrt{f(s)}\, ds] \tag{3.7.8}$$

As is evident from the denominator this analysis breaks down when f has a zero.

Consider next the case of f vanishing in the domain of interest. Specifically, we write

$$f(0) = 0,\ f'(0) \neq 0$$

since the zero can always be transformed to the origin. To be definite we take $f(x) > 0$, $x > 0$ and so $f(x) < 0$, $x < 0$. Then if u is some solution to (3.7.7) we can write

$$u \sim \frac{A}{f^{1/4}} \exp[i\lambda \int_0^x \sqrt{f(s)}\, ds] + \frac{B}{f^{1/4}} \exp[-i\lambda \int_0^x \sqrt{f(s)}]ds,\quad x > x_o > 0 \tag{3.7.9}$$

and

$$u \sim \frac{A'}{|f|^{1/4}} \exp[\lambda \int_0^x \sqrt{|f(s)|}\, ds] + \frac{B'}{|f|^{1/4}} \exp[-\lambda \int_0^x \sqrt{|f(s)|}\, ds], \quad x < -x_o < 0. \tag{3.7.10}$$

That is, for x bounded away from the origin we can still use the WKB solutions. Note that the second term in (3.7.10) is negligible. There are now two problems to study. First there is the connection problem, i.e., the relation between A,B and A',B'. Secondly, we wish to find the description of u in the neighborhood of the origin. This point, i.e., a zero of $f(x)$, for obvious reasons is called a transition point. (Due to reasons which we do not go into it is also called a turning point.)

We solve these problems by introducing a comparison problem. Certainly, the simplest ordinary differential equation incorporating all the features of our problem is

$$\frac{d^2w}{dx^3} - \lambda^2 xw = 0. \tag{3.7.11}$$

Setting $\zeta = \lambda^{2/3}x$ in (3.7.11) we get the Airy equation, whose solutions we studied in detail in Section 2.7. We recall the three solutions A_0, A_+, A_- (2.7.1) having the relation

$$A_0 + A_- + A_+ = 0 \tag{3.7.12}$$

and the asymptotic expansions,

$$\begin{aligned} A_0 &\sim \frac{\exp[-\frac{2}{3}\zeta^{3/2}]}{\zeta^{1/4}\sqrt{4\pi}}, \quad |\arg\zeta| < \pi \\ A_+ &\sim \frac{\exp[\frac{2}{3}\zeta^{3/2}]}{\zeta^{1/4}\sqrt{4\pi}}, \quad -\frac{\pi}{3} < \arg\zeta < \frac{5\pi}{3} \\ A_- &\sim \frac{-\exp[\frac{2}{3}\zeta^{3/2}]}{\zeta^{1/4}\sqrt{4\pi}}, \quad -\frac{5\pi}{3} < \arg\zeta < \frac{\pi}{3} \end{aligned} \tag{3.7.13}$$

We remark that $A_{\pm}$ are dominant on the real axis while A_0 is recessive there.

Therefore, for equation (3.7.11) the connection formulas follow from the results for the Airy integrals. To see this we first observe that (3.7.8) for f = x, gives (ignoring irrelevant constants)

$$\frac{\exp}{|x|^{1/4}}\left[\pm \frac{2}{3} i\lambda|x|^{3/2}\right], \quad x < 0$$

$$\frac{\exp[\pm \frac{2}{3} x^{3/2}]}{x^{1/4}}, \quad x > 0.$$

Also setting $\zeta = x\lambda^{2/3}$, x real, in the expression for the Airy functions (3.7.13)

$$(3.7.14) \qquad A_+ \sim \begin{cases} \dfrac{\exp[\frac{2}{3} \lambda x^{3/2}]}{\lambda^{1/6} x^{1/4} \sqrt{4\pi}}, & x > 0 \\[2ex] \dfrac{\exp[-\frac{2i}{3} \lambda|x|^{3/2} - \frac{i\pi}{4}]}{\lambda^{1/6}|x|^{1/4} \sqrt{4\pi}}, & x < 0 \end{cases}$$

$$(3.7.15) \qquad A_- \sim \begin{cases} -\dfrac{\exp[\frac{2}{3} \lambda x^{3/2}]}{\lambda^{1/6} x^{1/4} \sqrt{4\pi}}, & x > 0 \\[2ex] -\dfrac{\exp[\frac{2}{3} i\lambda|x|^{3/2} + \frac{i\pi}{4}]}{\lambda^{1/6}|x|^{1/4} \sqrt{4\pi}}, & x < 0 \end{cases}$$

$$(3.7.16) \qquad A_0(\lambda^{2/3}x) \sim \begin{cases} \dfrac{\exp[-\frac{2}{3} \lambda x^{3/2}]}{\lambda^{1/6} x^{1/4} \sqrt{4\pi}}, & x > 0 \\[2ex] \dfrac{2\cos(\frac{2}{3} \lambda|x|^{3/2} + \frac{\pi}{4})}{\lambda^{1/6}|x|^{1/4} \sqrt{4\pi}}, & x < 0. \end{cases}$$

Connection Formulas

Suppose u is a solution to (3.7.11) such that for $x < 0$

$$(3.7.17) \qquad u \sim \frac{A}{|x|^{1/4}} \exp[\tfrac{2}{3} i\lambda|x|^{3/2}] + \frac{B}{|x|^{1/4}} \exp[-\tfrac{2}{3} i\lambda|x|^{3/2}].$$

Then from (3.7.14) and (3.7.15) we have for $x > 0$

$$u \sim \frac{\left(Ae^{-\frac{i\pi}{4}} + Be^{\frac{i\pi}{4}}\right)}{x^{1/4}} \exp[\frac{2}{3} \lambda x^{3/2}]$$

Next suppose u has the property that for $x > 0$

$$u \sim \frac{B' \exp[-\frac{2}{3} \lambda x^{3/2}]}{x^{1/4}}$$

then from (3.7.12) and (3.7.16) we obtain for $x < 0$

$$u \sim \frac{2B' \cos(\frac{2}{3} |x|^{3/2} \frac{\pi}{4})}{|x|^{1/4}}$$

This summarizes the connection formulas. One situation has been deleted, i.e.,

$$u \sim \frac{A' \exp[\frac{2}{3} \lambda x^{3/2}]}{x^{1/4}}, \qquad x > 0.$$

All that can be said in this case is that for $x < 0$, u has the form (3.7.17); with A and B such that

$$A(e^{-i\pi/4} + Be^{i\pi/4}) = A'.$$

Stated in general terms if given a dominant expansion we cannot uniquely find the connection formulas. A little thought reveals that this of necessity must be the case.

The connection formulas which we found for (3.7.11) hold quite generally. For suppose we have (3.7.7) then we may approximate this equation by

$$\frac{d^2w}{dt^2} + \lambda^2 f'(0)xw = 0 \tag{3.7.18}$$

(recalling that $f(0) = 0$). Their setting

$$x = -(\operatorname{sgn} f'(0))|f'(0)|^{-1/3}y$$

we reduce (3.7.18) to the form (3.7.19). This, therefore, yields a description of the solution to (3.7.7) in the neighborhood of $x = 0$, the transition point. We note that the description is in terms of Airy functions.

To finish the argument we should note that solutions to (3.7.7) and (3.7.18) have an overlapping region. To see this we only have to expand (3.7.9,10) for $|x|$ small but bounded away from zero. This gives the same result as the WKB solutions of (3.7.18) under the same limit.

It should be clear that it was necessary to restrict attention to the real case. This is dealt with in the next exercise.

Exercise 88. Consider equation (3.7.7) with $f(0) = 0$, $f'(0) = C \neq 0$ and complex. Find the connection formulas. In particular, again point out the problem of connection to a dominant solution.

Exercise 89. Suppose in (3.7.7) that $f(0) = f'(0) = 0$, $f''(0) < 0$. Find the connection formulas. (First, find and solve a comparison problem.)

Langer's Uniform Method.

In the process of obtaining the connection formulas above, we obtained the solution of (3.7.7) in a neighborhood of the origin, i.e., where $f(x = 0) = 0$, and of course we also have the WKB solution for x bounded away from the origin. Langer [Trans. Am. Math. Soc. 33, 2-3, 1931; 34, 447 (1932); 36, 90 (1934). See also F.W.J. Olver, Phil. Trans. Roy. Soc. London, A 250: 60 (1958).] has introduced a method which at once describes both regions, i.e., a uniform method. This we now sketch.

We consider (3.7.7) under the conditions, $f(0) = 0$, $f'(0) \neq 0$, and introduce new variables

$$\xi = \varphi(x), \quad \eta = \psi(x)w.$$

Substituting into (3.7.7) we obtain

$$\eta_{\xi\xi} + \frac{1}{\varphi'}\left(\frac{\varphi''}{\varphi'} - \frac{2\psi'}{\psi}\right)\eta_\xi + \left(\frac{\lambda^2 f}{\varphi'^2} + \frac{\psi(\psi^{-1})''}{\varphi'^2}\right)\eta = 0.$$

Setting

$$\varphi' = \psi^2$$

we eliminate the coefficient of η_ξ. Next determining φ by taking

(3.7.19) $$\varphi^{3/2} = \frac{3}{2} i\lambda \int_0^x \sqrt{f(s)}\, ds$$

so that

$$\frac{\lambda^2 f}{\varphi'^2} = -\varphi$$

we obtain

(3.7.20) $$\eta_{\xi\xi} - \xi\eta = \frac{1}{\lambda^{4/3}} r(\xi)\eta.$$

The coefficient $r(\xi)$ is determined by

$$r(\xi) = -\frac{3}{4}\frac{\lambda^{4/3}\varphi''}{(\varphi')^4} + \frac{1}{2}\frac{\varphi'''\lambda^{4/3}}{(\varphi')^3}$$

and from (3.7.19) we also see that r is independent of λ. Also from (3.7.19) we see that $\varphi = O(x)$ for $x \sim 0$ and, therefore, $\varphi'(0) \neq 0$, from which it follows that $r(\xi)$ is bounded. Neglecting the right hand side of (3.7.20) in the limit $\lambda \to 0$ we see that approximate solutions are given by Airy functions,

$$\eta \sim A_i(\xi)$$

where A_i is used to symbolize the Airy functions. Then since

$$\varphi = \left[\frac{3}{2}\, i\lambda \int_0^x \sqrt{f(s)}\, ds\right]^{2/3}$$

and

$$\psi = \left[-\frac{\lambda^2 f}{[\frac{3}{2}\, i\lambda \int_0^x \sqrt{f(s)}\, ds]^{2/3}}\right]^{1/4}$$

we can write the uniformly valid asymptotic solution to (3.7.7) as

$$w \sim \frac{(\int_0^x (f(s))^{1/2} ds)^{1/6}}{f^{1/2}} \Big\{ \alpha A_i([\tfrac{3}{2}\, i\lambda \int_0^x (f(s))^{1/2} ds]^{2/3})$$

(3.7.21)

$$+ \beta B_i([\tfrac{3}{2}\, i\lambda \int_0^x (f(s))^{1/2} ds]^{2/3}) \Big\}$$

where A_i and B_i denote two linearly independent solutions of Airy's equation and α, β are constants.

Exercise 90. Using (3.7.21), obtain the connection formulas.

Exercise 91. Consider (3.7.7) with $f(0) \neq 0$, i.e., without a turning point, and find a transformation which reduces it to

$$\frac{d^2\eta}{d\xi^2} + \eta = \frac{1}{\lambda^2}\, \rho(\xi).$$

Obtain the WKB solutions from this.

A problem which we do not consider here is that of an ordinary differential equation containing many transition points. For example, the parabolic cylinder function differential equation (3.0.15) is canonical for the case of two transition points. The general treatment of a second order equation containing many transition points has been recently given in [R. Lynn and J. B. Keller, CPAM 23, 3(1970).]

The explanation of the letters WKB has been deliberately avoided, since this opens an elaborate and involved history of priorities. An account of this is given in [J. C. Heading, An Introduction to Phase-Integral Methods, John Wiley (1962)]. This book contains an exhaustive treatment of the subject as well as a number of applications. A more recent account is given in [N. and P. Frohman, JWKB Approximation, North Holland Press, Amsterdam, 1965]. For other sources see [2], [6], [7], [8].

Reading List

[1] K.O. Friedrichs, Special Topics in Analysis, New York University Notes.

[2] A. Erdelyi, Asymptotic Expansions, Dover Publications, New York, 1956.

[3] E. Copson, Asymptotic Expansions, Cambridge University Press, 1965.

[4] N. de Bruijn, Asymptotic Methods in Analysis, North Holland Press, Amsterdam, 1958.

[5] H. Lauwerier, Asymptotic Expansions, Math. Centrum (Holland).

[6] H. Jeffrys, Asymptotic Approximations, Oxford University Press, 1962.

[7] W. Wasow, Asymptotic Expansions for Ordinary Differential Equations, John Wiley, New York, 1965.

[8] G. Carrier, M. Krook, C. Pearson, Functions of a Complex Variable, McGraw-Hill, New York, 1966.

[9] G. Carrier and C. Pearson, Ordinary Differential Equations, Ginn Blaisdell, 1968.

[10] E. Coddington and N. Levinson, Theory of Ordinary Differential Equations, McGraw-Hill, New York, 1955.

[11] R. Bellman, Perturbation Techniques in Mathematics, Physics, Engineering, Holt, 1964.

[12] J. Cole, Perturbation Techniques in Applied Mathematics, Blaisdell, Waltham, Massachusetts, 1968.

[13] J. Dieudonne, Calcul Infinitesimal, Hermann, Paris, 1968.

[14] W. Magnus, F. Oberhettiger, R. Soni, Formulas and Theorems for the Special Functions of Mathematical Physics, Springer-Verlag, New York, 1966.

[15] M. Abramowitz and I.A. Stegun, Handbook of Mathematical Functions, U.S. Government Printing Office, Washington, 1964.

[16] E. Ince, Ordinary Differential Equations, Dover Publications, 1956.

INDEX

Airy's equations, 127, 191, 289, 293

Airy's functions, 126, 174

Airy's integrals, 126, 134, 171, 174

relation to transition points, 293

Stoke's lines of, 133, 134

analytic continuations, 3, 68, 129

of functions defined by an integral, 38, 39

of Laplace transforms, 69

asymptotic developments, 4, 7

error estimates, 25, 60

extended sense, 10, 11

general, 8

in the sense of Poincare, 9

uniqueness, 10, 12, 13

asymptotic expansions, 11, 24, 26

of Airy's integrals, 132, 133, 288, 293

of Bessel functions, 126, 288

of Bromwich integrals, 182, 186

of Fourier integrals, 74, 181

of Laplace integrals, 66, 71, 72, 78, 274

asymptotic integration, 17, 19

asymptotic power series, 2

derivative of, 18, 21

uniqueness, 12

asymptotic sequences, 7

Bellman, R., 300

Bessel functions, 9, 122, 126, 170

reduction to Airy's function, 171

asymptotic expansions of, 126

Bessel's equations, 287

Bleistein, N., 140, 170, 171

Borel sum, 35

Bromwich integral, 182, 186

Bromwich path, 183

Carrier, G., 300

Cauchy integral formula, 23, 210

Cauchy sequence, 207

Cayley-Hamilton theorem, 208

Chako, N., 140

characteristic polynomial of a matrix, 202, 209

Chester, 171, 173

circuit relation, 230

classical adjoint of a matrix, 193, 209

Coddington, E., 188, 300

cofactor matrix, 193

Cole, J., 300

connection formulas for transition points, 294, 298

convolution integrals, 157

Copson, E. T., 300

critical points, 88, 93, 137, 147

de Bruijn, N., 300

derivatives of APS, 18, 21

Dieudonne, J., 10, 44, 300

dispersive wave, 102

dominant solution, 270, 295

Dunford-Taylor integral, 210, 217, 235, 238

edge, 270

eigenvalues, 201

    degenerate, 204, 265

    multiplicity of, 202, 225, 235, 242

eigenvectors, 201

    generalized, 239, 242, 244, 246

Erdelyi, A.,10, 76, 79, 300

error estimate of AD, 25, 60

error functions, complementary, 73, 161

essential singularities, 3, 229, 232, 258, 260

Euler transformation, 29, 33

Euler's constant, 177

Euler's method, 29

existence of solution for ODE, 188

exponential function of a matrix, 215, 216

Focke, J., 137, 140

Fourier integrals, 62, 74, 88, 157, 163, 181

Friedman, B., 171, 173

Friedrichs, K. O., 178, 300

Fróbenius, method of, 255

Frohman, 299

Fuch's theorem, 232, 241, 255

Fuch's type, equation of, 258

functions of matrices, 207, 209, 211

fundamental matrix solution, 196, 200, 230, 269, 271

fundamental system of solutions, 196

Gel'fand, I. M., 139

Gamma function, 25

gauge functions, 8, 9, 17, 18

group velocity, 102

Halmos, P., 192

Hamburger equation, 288

Handelsman, R., 140, 178

Hankel function, 73, 122

Heading, J. C., 299

Heaviside function, 168

Hermitian matrix, 193, 203

Hilbert transform, 182

Huo, Wei-chi C., 104

hypergeometric equation, 257, 258

hypergeometric functions, 159

Ince, E. L., 283

indefinite integrals, asymptotic evaluation of, 14

indicial equation, 256, 257

integral representations of

    Airy's function, 126

    Bessel functions, 126

    Gamma functions, 25

    Hankel functions, 73, 122

integration by parts, 40, 164

irregular singular point, 230, 234, 259, 278

Jeffrys, H., 300

Jones, D. S., 140

Jordan canonical form, 204

Keller, J., 299

Kelvin's formula, 86, 87, 101, 104

generalized, 100

multidimensional integral, 136

Kline, M., 140

Krook, M., 300

Landau symbols, 5, 6, 7

Langer, R., 296

Laplace integrals, 66, 71, 72, 80, 88, 114, 274

Laplace transforms, 62, 70, 77, 80, 176, 182, 184

Laplace's formula, 80, 83, 96, 164

Laplace's method, 80, 272

Laurent expansions, 286

Lauwerier, H., 299

level curves, 106

Levinson, N., 188, 300

Lew, J., 178

Lynn, R., 299

Magnus, 122, 160, 166

matrix, 192, 195, 201

adjoint of, 193

canonical forms, 204

characteristic polynomials, 202, 209, 214

classical adjoint of, 193

cofactor, 193

diagonal, 202

function of, 207, 211

inner products, 192

Hermitian, 193, 203

Jordan canonical form of, 204

minimal polynomials, 212

normal, 193, 203

norms of, 205, 206

null space of, 194

ranks, 194

similarity, 203, 230

trace, 198

matrix solutions, fundamental, 196, 230, 269, 271

minimal polynomials, 212, 214

multiplicity of eigenvalues, 202, 225

neutralizers, 75, 88, 93, 138, 140, 142

Newton's polygon, 280

normal matrix, 193

normal solution, 271

null space, 194

Oberhettinger, F., 122, 160, 166

Olver, F. W. F., 27, 296

orthogonal transformations, 141, 144, 203, 230

parabolic cylinder equation, 191

parabolic cylinder function, 160, 161, 166

Pearson, C., 300

Poincare, H., 13

polynomials, characteristic, 202, 209

minimal, 212

ranks of matrices, 194

recessive solutions, 270

regular singular points, 230, 234, 254, 283

Riccati's equation, 190

Riemann-Lebesgue theorem, 63

generalized, 64

Riemann notation, 258

Ritt's theorem, 36

saddle, monkey, 115, 171

saddle points, 105, 113, 117, 126

coalescing of, 171

formula, 105, 121, 126

formula for complex large parameter, 115

hills of, 113

valleys of, 113, 174

scaler ordinary differential equations, 234, 254, 278, 285

definition of irregular singular points, 278

definition of regular singular points, 234, 254, 257, 278

Shanks, D., 35

shearing transform, 240

Shilov, G., 139

singular points of ODE, 226, 232

irregular, 230, 234, 259, 278

regular, 230, 234, 254, 257, 273, 276

Sirovich, L., 104, 162

Soni, R., 122, 160, 166

stationary phase, method of, 86

path, 106

stationary point, 87, 99, 100

steepest descent, method of, 105

path, 106, 107, 108

Stirling's formula, 84

Stokes lines, 3, 60, 73, 133, 134, 270, 275

subdominant solution, 270

subnormal solution, 271

Taylor's theorem, 4

transition point, 292, 297, 298, 299

turning point, 297

uniform asymptotic expansion, 164-175, 296

uniqueness of solution for ODE, 188, 189

Ursell, 171, 173

van der Corput, 75

variation of parameter method, 200

Wasow, W., 289, 300

Watson's lemma, 65, 66, 70

WKB method, 126, 291, 292, 296, 299

Wilcox, C. H., 27

Wronskian, 197